Ocular Cytopathology

Ocular Cytopathology

Ben J. Glasgow, M.D.
Assistant Professor of Ophthalmology and Pathology
UCLA School of Medicine
Jules Stein Eye Institute
Los Angeles, California

Robert Y. Foos, D.V.M., M.D.
Professor of Pathology
UCLA School of Medicine
Jules Stein Eye Institute
Los Angeles, California

Butterworth–Heinemann
Boston London Oxford Singapore Sydney Toronto Wellington

Library of Congress Cataloging-in-Publication Data
Glasgow, Ben J.
Ocular cytopathology / Ben J. Glasgow, Robert Y. Foos.
p. cm.
Includes bibliographical references and index.
ISBN 0-7506-9260-X :
1. Eye—Cytopathology. I. Foos, Robert Y., 1922– .
II. Title.
[DNLM: 1. Biopsy, Needle—methods—atlases.
4. Orbital Diseases—pathology—atlases. WW 17 G54o]
RE66/5/G58 1992
DNLM/DLC
for Library of Congress 92-19272
CIP

British Library Cataloguing-in-Publication Data
A catalogue record for this book is available from the British Library.

Butterworth–Heinemann
80 Montvale Avenue
Stoneham, MA 02180

10 9 8 7 6 5 4 3 2 1

Printed in the United States of America

To Carol and Sunny

Contents

Preface

The aim of this book is to provide pathologists, ophthalmologists, and students of opthalmic pathology with information that will facilitate the interpretation of ocular specimens derived from exfoliative cytology, intraocular washings, and fine needle aspiration of the eye and its adnexa. There is emphasis on techniques used to obtain and process laboratory specimens. Clinical and radiologic appearances are stressed where relevant to cytologic interpretation.

This book is not intended to be a detailed exposition of all ocular diseases; rather, it will serve as a practical guide with information needed to interpret common eye cytology specimens. It is written for the general pathologist or cytologist with little in-depth training in eye disease who is asked to interpret ocular specimens. Abundant clinical, radiologic, and histologic material is presented to provide a general perspective for those with little previous experience in ophthalmology.

Hopefully, this book will be a valuable resource for clinicians faced with diagnostic problems that may be solved by cytologic techniques. In addition, detailed information regarding processing will assist clinicians in choosing appropriate investigative techniques.

This book will help students of ophthalmic pathology learn the basics of eye cytology and apply knowledge of histology to cytologic features of eye diseases. To aid in this transition, numerous gross and microscopic color photographs are presented.

Acknowledgments

We extend our gratitude to our department chairmen, Bradley R. Straatsma (ophthalmology) and Pasquale A. Cancilla (pathology), for their continuing support of the ophthalmic pathology program at the Jules Stein Eye Institute, UCLA School of Medicine. We are grateful to the many clinical colleagues who shared their case material with us, especially Thomas H. Pettit, Howard R. Krauss, Robert A. Goldberg, Michael J. Groth, Lynn Gordon, James D. Shuler, Gary N. Holland, and Allan E. Kreiger without whom this book would not have been possible.

CHAPTER

1 Methods in Ocular Cytology

In this chapter, methods for processing ocular cytology specimens are described. Most of the routine eye cytology specimens can be prepared by simple fixation and staining methods; although, occasionally more elaborate techniques, such as electron microscopy and immunocytochemistry, are helpful.

FIXATION AND STAINING METHODS

Each fixative and stain has advantages and disadvantages. In general, slides are either rapidly air dried or rapidly fixed in 95% ethanol. Air drying artificially expands the cells while ethanol artificially shrinks the cells. If air dried, a modified Wright or Giemsa stain should be used.[1] May-Grünwald Giemsa demonstrates excellent cytologic differentiation and is especially good for cells of hematopoietic origin. The Papanicolaou stain is excellent for squamous lesions. Hematoxylin and eosin recapitulates standard histopathology and is often preferred by those adept at surgical pathology, while Papanicolaou is preferred by those expert in exfoliative cytology. A combination of stains is helpful. The method of fixation and staining is determined by the clinical question to be answered. If allergic conjunctivitis is to be differentiated from infectious conjunctivitis, then air-dried, Giemsa-stained preparations are appropriate to differentiate eosinophils and neutrophils. If squamous dysplasia or carcinoma is suspected, then ethanol-fixed Papanicolaou stains are most useful. For suspected chlamydia conjunctivitis, ethanol-fixed material reacted with direct fluorescent antibody is very sensitive.[2] Giemsa-stained, air-dried smears can be done as an adjunctive procedure. Example protocols for staining by May-Grünwald Giemsa, Papanicolaou, and hematoxylin and eosin are shown in Tables 1–1, 1–2, and 1–3.

TABLE 1–1 **Protocol for May-Grünwald Giemsa staining**

Solution	*Time/Procedure*
Methyl alcohol	5 min
May-Grünwald	10 min
Tap Water	10 s
Giemsa	30 min
Tap water	10 s
Dry slide	~10 min
Dip xylene	rapid dip
Mounting media	cover slip

CONJUNCTIVAL AND CORNEAL SMEARS

Scrapings from the conjunctiva and cornea may be performed with a platinum spatula, cotton swab, or cytology brush (Figure 1–1).[3–5] The platinum spatula is easily sterilized, reuseable, and readily available in most ophthalmology offices and clinics. Swabs may leave cotton fibers on the slide that can confuse and annoy the microscopist. In addition, cotton swabs have the disadvantage of adhering to cells so that fewer are transferred to slides. However, cytologic smears made by rolling or smearing the cotton swab on the slide can be quite adequate. Immediately after scraping or swabbing the conjunctiva or cornea, the cells are rapidly smeared on a

Table 1–2 Protocol for Papanicolaou Staining

Solution	*Time/Procedure*
70% ethanol	10 dips
Distilled water	10 dips
Mayer's hematoxylin	3 min
Tap water	7 min
70% ethanol	10 dips
95% ethanol	10 dips
95% ethanol	10 dips
Orange G	1 min
95% ethanol	10 dips
95% ethanol	10 dips
Eosin azure	1.5 min
95% ethanol	10 dips
95% ethanol	10 dips
Absolute ethanol	10 dips
Absolute ethanol	10 dips
Absolute ethanol:xylol (1:1)	1 min
Xylol	10 dips
Xylol	10 dips
Xylol	10 dips
Mounting media	cover slip

Table 1–3 Protocol for Hematoxylin and Eosin Rapid Staining

Solution	*Procedure/Time*
95% ethanol	Fixative and storage
Distilled water	10 dips
Mayer's hematoxylin	1 min
Acid alcohol	1 dip
Tap water	5 dips
Lithium carbonate	Until blue
Distilled water	5 dips
80% ethanol	3 dips
Eosin	1 min
95% ethanol	5 dips
95% ethanol	5 dips
100% ethanol	5 dips
Xylene	2 to 3 dips
Xylene	2 to 3 dips
Mounting media	cover slip

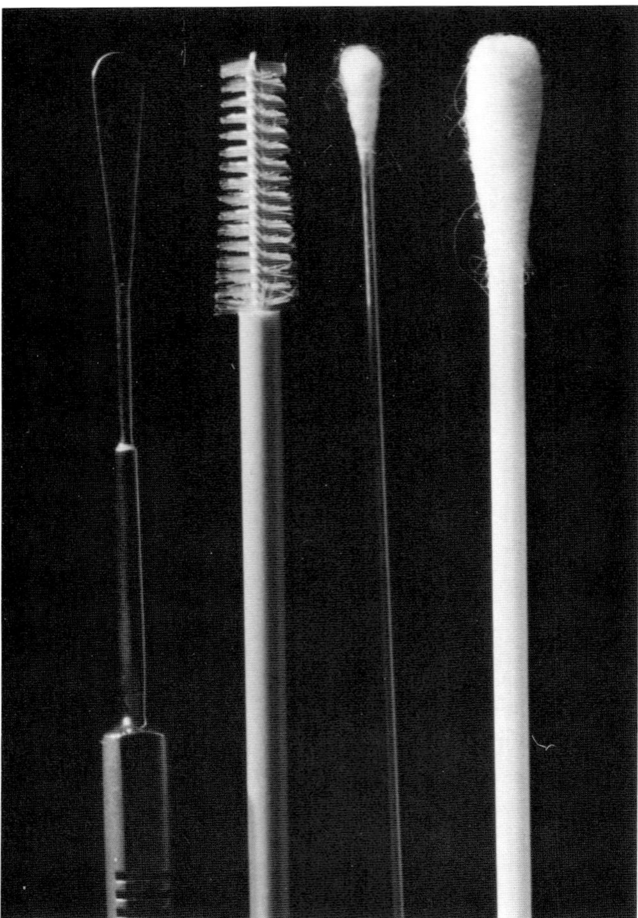

FIGURE 1–1 A variety of instruments can be used to smear conjunctival and corneal cells, including a platinum spatula, cotton swab, or cytology brush.

glass microscope slide and instantly dropped into a cytology fixative, such as 95% ethanol. It is important at this stage to fix the cells immediately so that they are not allowed to air dry. Rapidly ethanol-fixed preparations can be stained by Papanicolaou or hematoxylin and eosin methods. For air-dried material, slides must be rapidly dried to avoid cellular explosion artifacts. The air-dried slides can be stored for several hours, fixed with methanol, and stained with May-Grünwald Giemsa, as described above.

Conjunctival impression cytology is performed by examining cellulose acetate filter preparations that have been placed against the conjunctiva to remove cells.[6] The choice of pore size in the filter paper affects both morphology and cellular yield.[7] It has proven useful in the study of keratoconjunctivitis sicca,[8] vitamin A deficiency,[9, 10] cicatricial ocular pemphigoid,[11] analysis of inflammatory cell composition in allergic and infectious conjunctivitis,[12] identification of neonatal chlamydia infection,[13] and squamous metaplasia from any cause.[14, 15] The technique has been modified for electron microscopy to identify mucopolysaccharidoses.[16] The impression method has advantages compared to scraping techniques. It is simple to perform, is atraumatic, and can be used to topographically map the extent of conjunctival lesions. In less-experienced hands, impression cytology has a higher cellular yield than use of either cotton swabs or a cytology brush.[17] Impression cytology is currently employed mainly as a research tool.

PARS PLANA VITRECTOMY

Most intraocular cytology specimens are obtained from pars plana vitrectomy. On average, 250 vitrectomy specimens are processed each year at the Jules Stein Eye Institute. Routine processing of all vitrectomy specimens can be helpful because unexpected lesions may be identified, such as amyloidosis, lymphoma, and inadvertent retinectomy. It is important that the cytologist understand how these specimens are obtained and handled before they reach the laboratory. Retinal surgeons perform vitrectomies for a variety of indications ranging from diagnosis of infection to elimination of vitreous traction on the retina. In general, very small incisions are made in the pars plana region of the eye for the light port, vitrectomy cutting instrument, and infusion cannula. The vitrectomy instrument provides suction to aspirate the vitreous while, at the same time, cutting. A fiber-optic illuminator is introduced to allow direct lighting of structures of the posterior segment of the eye during the operation. The infusion cannula provides a flow of balanced salt solution or gas to keep the eye expanded (Figures 1–2 and 1–3). Material that is aspirated by the vitrectomy instrument travels through plastic tubing to the collection bag. Thus, specimens from the eye have been diluted and are not in fixative when they arrive at the laboratory. Accordingly, some method of concentrating or filtering the specimen is necessary in processing. Specimens should be refrigerated immediately upon removal and processed promptly.

METHODS FOR ANALYZING PARS PLANA VITRECTOMY SPECIMENS

Vitrectomy specimens have been processed by a variety of techniques, including celloidin bag techniques,[18] cytocentrifugation,[19] direct smears, and millipore filtration.[20]

Some authors have compared these techniques and have found that cytocentrifugation revealed better cellular recovery and cytologic detail, but that the celloidin bag specimens more closely resembled histopathologic patterns.[21] Others found that recovery rates and cellular morphology were superior with the millipore filtration method as compared to the cytocentrifuge technique.[22] Cytocentrifugation is generally performed in two stages. Because the vitrectomy specimen is dilute, a preliminary centrifugation step is necessary. Optimal cellular sedimentation occurs at about 600 gravity, centrifugal (g) (about 1000 revolutions per minute (rpm) for 10 min on most cytocentrifuges, but calibration is necessary).[23] The pellet is transferred to an angled plastic tube that is separated from a glass slide by filter paper and tightly clipped (Figure 1–4). Centrifugation can be done with several types of centrifuges, such as the Shandon 2 cyto-centrifuge (Shandon, Pittsburgh, PA) or the Damon IEC (Interna-

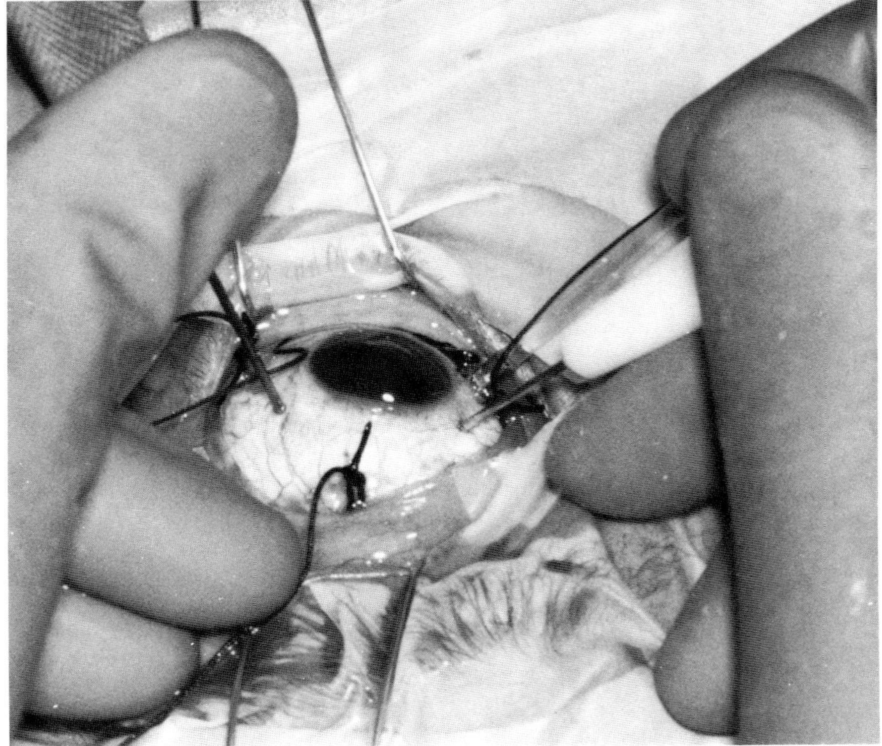

FIGURE 1–2 Clinical photograph of vitrectomy in progress with fiber-optic light (lower right), infusion cannula (upper right, and vitrectomy cutting instrument (left).

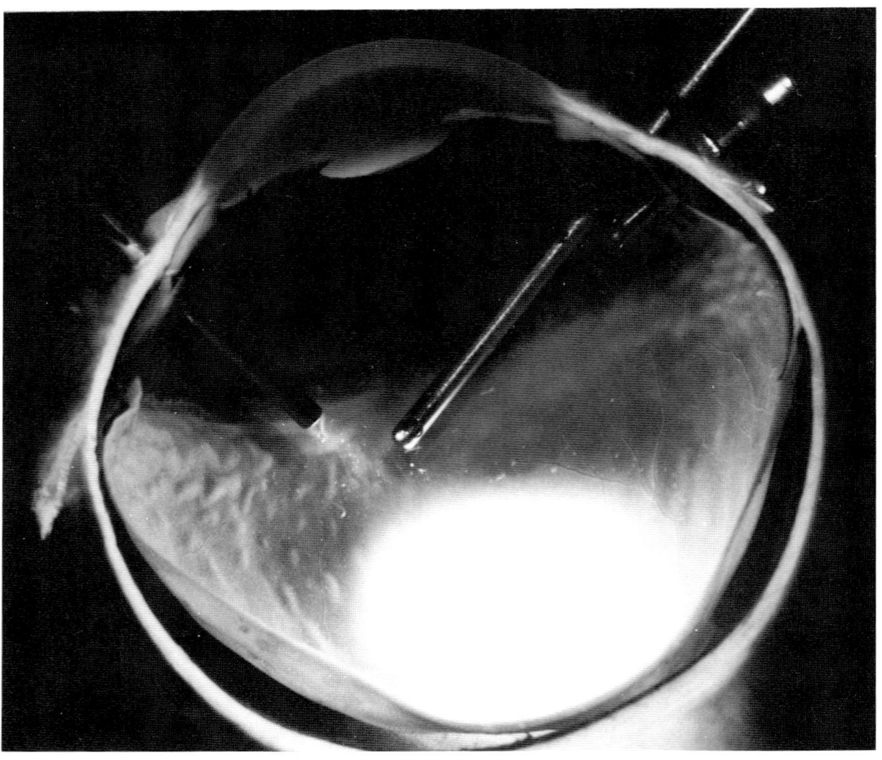

FIGURE 1–3 Photograph of a sectioned eye illustrates the infusion cannula (lower right), fiberoptic light pipe (left), and vitrectomy cutter (upper left) during vitrectomy.

FIGURE 1–4 The arrangement of slide, filter paper, and plastic centrifuge tube for processing vitreous washings is shown disassembled.

tional Equipment Company, Needham Heights, MA). Calibration is necessary to attain about 600 g (e.g., 650 rpm) for 5 min. These steps concentrate cells on the glass slide. Fixation is immediately performed with either rapid air drying or alcohol fixation. With cytocentrifugation, cells may be adsorbed to the filter paper and not recovered for cytologic study. It is, however, a very time efficient method for rapid sampling of intraocular washings.[24]

Millipore filters use a 46-millimeter (mm) diameter, 5 μm pore-size filter in which vacuum assists filtration of fresh vitreous washings. The cells are fixed by passing 95% ethyl alcohol through the filter.[16] Because the cells are continually submerged in fixation solutions, there are numerous steps that reduce cellular recovery.[25]

The celloidin bag method is another technique for cell collection.[16] Vitreous specimens are allowed to settle in syringes, combined with formalin, and placed into a bag of celloidin previously solidified in the shape of a test tube. The specimens are centrifuged for 10 min, then the celloidin bags are removed and placed in a cassette for processing by standard paraffin embedding techniques. Cells may escape if the bag ruptures before or during paraffin processing.

ELECTRON MICROSCOPY

Electron microscopy of small cytology samples (i.e., vitrectomy washing and fine needle aspirates) requires precise processing technique. The decision to perform electron microscopy on a given cytology specimen is best made prior to the operation or procedure to obtain the cells. Fixative selection and embedding compounds are determined by differential diagnoses generated from clinical and radiologic information in each case. Cytologic examination of the first fine needle aspirate is often critical to deciding on appropriate processing methods for the other aspirates. The mixture of 2% paraformaldehyde and 2% glutaraldehyde fixative is adequate fixation for discerning fine ultrastructural detail. Four percent paraformaldehyde fixative and a porous embedding medium can be used for immunoelectronmicroscopy.[26] Centrifugation of the sample after each step, followed by hand pipetting of solutions, improves the cellular yield (Tables 1–4 and 1–5). In this way, the specimen may be completely processed in a small polyethylene microfuge tube (Figure 1–5). The tube may be cut away from the polymerized epon-embedded specimen leaving the block from which to cut sections. Many techniques are available for fixing and staining for electron microscopy.[27]

IMMUNOCYTOCHEMISTRY

Immunocytochemistry can be performed quite easily on ocular cytology specimens. Alcohol fixation has been shown to be superior to formalin for preserving cellular antigens for cytology.[28] However, one must consider the type of antigen to be detected in choosing the appropriate fixative. Alcohol fixation is adequate for immunocytochemistry of common cytoplasmic antigens, such as those in melanoma proteins, glial proteins, muscle proteins, and cytokeratins. Air-dried tissue may also be used effectively for immunocytochemistry.

TABLE 1–4 An Embedding Protocol for Electron Microscopy

1. Specimen is placed in a microfuge tube with appropriate fixative (2% paraformaldehyde and 2% glutaraldehyde in 0.1 molar (M) sodium cacodylate buffer).
2. Centrifugation in microfuge for 4 min.*
3. Pipette rinse of 0.1 M sodium cacodylate buffer.
4. Osmicate one h in 1% osmium tetroxide 0.1 M sodium cacodylate buffer.
5. Rinse in 0.1 M sodium cacodylate buffer.
6. Dehydrate in dimethoxypropane and acetone (1:2) for 10 min to 1 h.
7. Rinse in acetone 1 to 2 min.
8. Pipette into mixture of 50% epon and 50% acetone for 1 h.
9. Pipette 100% epon** from 2 to 12 h.
10. Polymerize in fresh epon at 60°to 70°C

*Centrifugation should be performed to pellet the cells prior to each solution change.

**Epon refers to an epoxy resin that contains appropriate polymerization agents.

TABLE 1–5 Embedding Protocol for Immunoelectron Microscopy

1. Specimen is placed in a polyethylene microfuge tube with appropriate fixative (4% paraformaldehyde in 0.2 Molar (M) sodium phosphate buffer).
2. Centrifugation in microfuge for 4 min and prior to each solution change.
3. Pipette rinse with 0.2 M sodium phosphate buffer.
4. Rinse in 50% methanol for 30 min.
5. Pipette in 1% uranyl acetate and 70% methanol for 1 hr.
6. Rinse in 70% methanol x 2.
7. Place in LR white and 70% methanol 2:1 for 6 h.
8. Place in LR white overnight.

Figure 1–5 A microfuge tube facilitates processing of cytology specimens for electron microscopy.

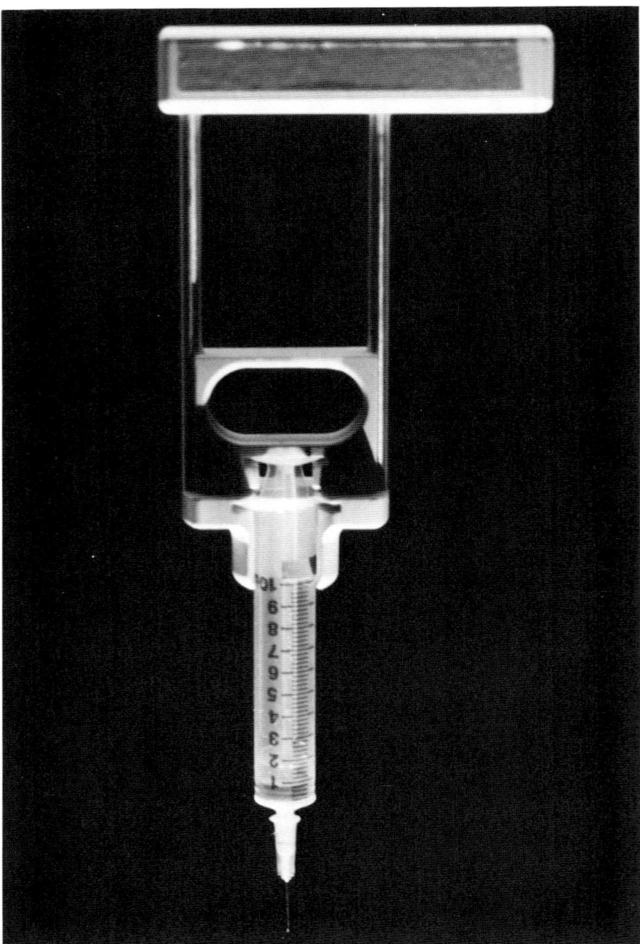

Figure 1–6 Photograph of arrangement for fine needle aspiration using the aspiration gun syringe directly connected to the needle.

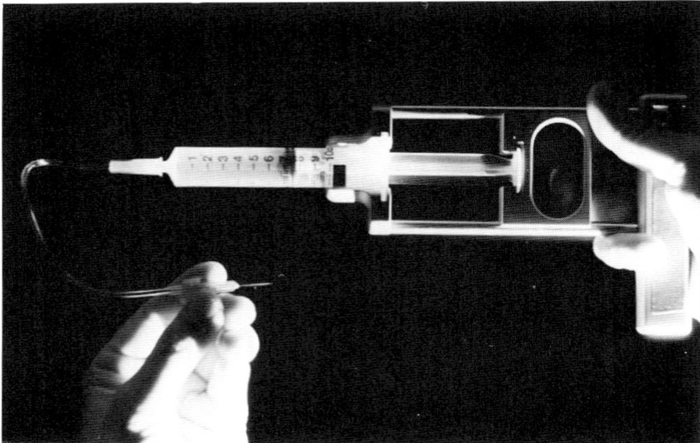

Figure 1–7 Extension tubing placed between the syringe and the needle allows the aspirationist to control the needle with maximum dexterity.

The protocols for immunocytochemistry performed on cytology specimens include incubation with an organic solvent step (usually acetone) to allow cellular penetration of the antibody and reaction with cytoplasmic antigens.[29] Otherwise, the protocols are very similar to those used in tissue pathology. An appropriate fixative for lymphomas is ethanol,[30] although some laboratories prefer to collect the cells in fresh buffered media, gently centrifuge, and rapidly fix in alcohol.

FINE NEEDLE BIOPSY

The principle of fine needle (aspiration) biopsy is that the sharp bevel of the needle cuts small pieces of tumor tissue that can be drawn into the needle by aspiration or capillary action. The material is then smeared on slides and fixed by one of the methods described earlier. There are several variations to the technique. A gun syringe holder, a 10 cubic centimeter (cc) syringe, and a needle (23 to 30 gauge) may be used to obtain the sample (Figure 1–6). However, for orbital lesions and intraocular lesions, fingertip control can be achieved by using sterile intravenous extension connecting tubing between the needle and the syringe (Figure 1–7). Biopsies can be done with a small needle placed directly into the tumor without aspiration (Figure 1–8).[31] A slight to-and-fro movement of the needle will usually enable a better sampling of material, but this movement may need to be restricted according to the size of the tumor and the anatomic site. A scant, but diagnostic sample can often be obtained without any movement.

The placement of the needle in orbital and ocular tumors requires precise localization. For nonpalpable orbital lesions, localization can be guided using computerized tomography (CT) or ultrasound. Intraocular tumors can be aspirated under direct visualization using a contact lens and an operating microscope. Aspirates can be performed by a direct or indirect approach to the tumor. The direct technique involves placing the needle through the adjacent sclera and into the tumor. In the indirect technique, the needle is inserted through the pars plana and intervening vitreous into the choroidal tumor on the opposite side (Figure 1–9). For iris or ciliary body tumors, the needle can be placed through the corneal limbus and aqueous to reach the lesion. The choice of needle size is important in intraocular aspirates. Adequate sampling of intraocular tumors with 30 gauge needles has been obtained both in bench specimens and in patients.[32,33] In our experience, the smaller the needle size, the less hemorrhage and traumatic injury will be induced. Only rarely is a large needle helpful in obtaining an adequate specimen.

Tumor implantation metastases from needle aspiration biopsy have been documented, but not from cases in which needles of 25 gauge or less have been used. Preliminary evidence suggests that too few cells are seeded for implantation metastases to occur with this method.[34]

The technique for aspirating tissues and making cy-

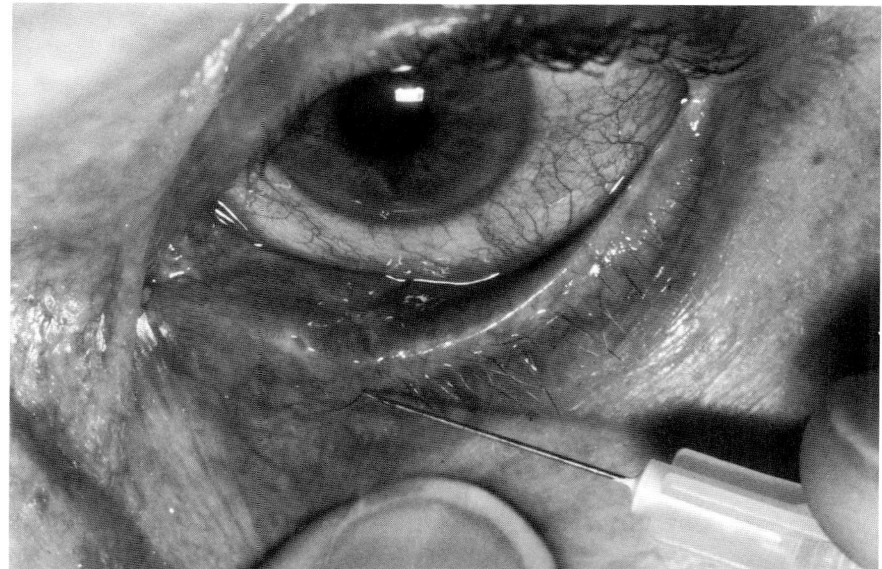

FIGURE 1–8 Fine needle biopsy of an eyelid lesion without aspiration.

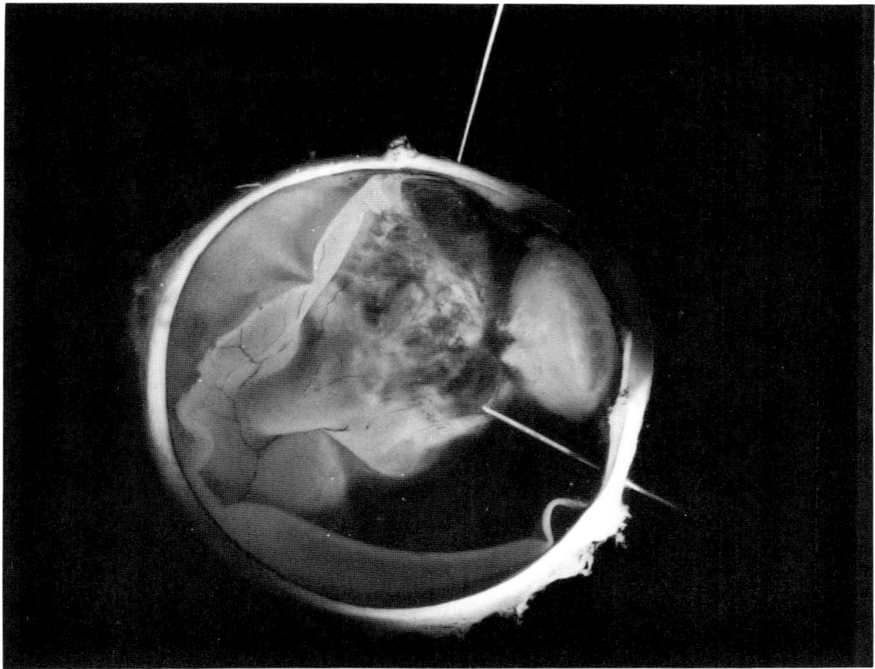

FIGURE 1–9 The direct and indirect methods of fine needle aspiration are illustrated in a bisected surgical specimen.

tology smears is very simple, but it requires practice to consistently obtain diagnostic material and produce evenly spread tissue preparations. Mistakes in procuring and smearing samples are a frequent cause of inadequate or uninterpretable results. The simplest method of smearing is to use two glass slides, one perpendicular to the other in the horizontal plane and slightly angled in the vertical plane. The angled slide is gradually flattened against the tissue on the other slide while pulling smoothly downward (Figure 1–10). For the clinician who rarely performs this procedure, it is best to seek the advice of an experienced pathologist for smear preparation, fixative and stain selection, and processing.

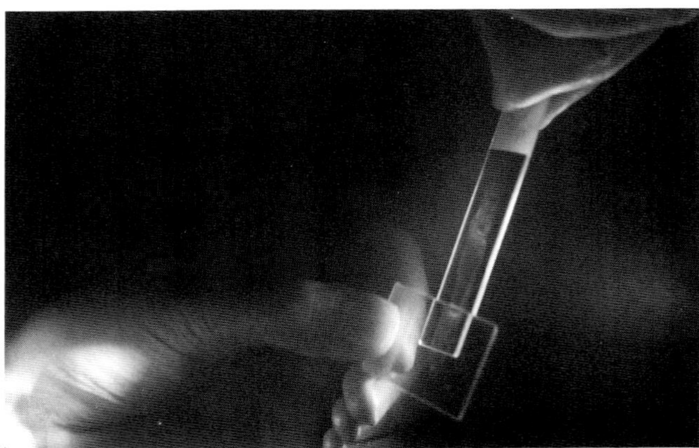

FIGURE 1–10 **One technique of smearing aspirated tissue fragments between two glass slides.**

REFERENCES

1. Reich C. Modified Wright stain. Am J Clin Pathol 1954;24:881.

2. Bell TA, Kuo C, Stamm WE, Tam MR, Stephens RS, et al. Direct fluorescent monoclonal antibody stain for rapid detection of infant *Chlamydia trachomatis* infections. Pediatrics 1984;74:224–228.

3. Thygeson P. The cytology of conjunctival exudates. Am J Ophthalmol 1946;29:1499–1512.

4. Kimura S, Thygeson P. The cytology of external disease. Am J Ophthalmol 1955;39:137–145.

5. Duszynski L. Cytology of the conjunctival sac. Am J Ophthalmol 1954;37:576–578.

6. Egbert PR, Lauber S, Maurice DM. A simple conjunctival biopsy. Am J Ophthalmol 1977;84:798–801.

7. Martinez JA, Yee RW, Tio FO. Improving the standardization of conjunctival impression cytology. Invest Ophthalmol Vis Sci 1991;32:734.

8. Nelson JD, Havener VR, Cameron JD. Cellulose acetate impressions of the ocular surface. Arch Ophthalmol 1983; 101:1869–1872.

9. Wittpenn JR, Tseng SCG, Sommer A. Detection of early xerophthalmia by impression cytology. Arch Ophthalmol 1986;104:237–239.

10. Natadisastra G, Wittpenn JR, West KP, Muhilal RD, Sommer A. Impression of cytology for detection of vitamin A deficiency. Arch Ophthalmol 1987;1224–1228.

11. Nelson JD. Ocular surface impression using cellulose acetate filter material: ocular pemphigoid. Surv Ophthalmol 1982;27:67–69.

12. Thatcher RW, Darougar S, Jones BR. Conjunctival impression cytology. Arch Ophthalmol 1977;95:678–681.

13. Maskin SL, Heitman KF, Yee RW. Use of impression cytology in neonatal chlamydial conjunctivitis. Arch Ophthalmol 1987;105:1626.

14. Tseng SCG. Staging of conjunctival squamous metaplasia by impression cytology. Ophthalmology 1985; 92:728–733.

15. Grene RB, Lankston P. Cartography of impression cytology. Cornea 1990;9:275–278.

16. Maskin SL, Bode DD. Electron microscopy of impression-acquired conjunctival epithelial cells. Ophthalmology 1986;93:1518–1523.

17. Connor CG, McGinn TT, Tirey WW. Comparison of cytologic techniques used in differential diagnosis of conjunctivitis. Invest Ophthalmol Vis Sci 1991;32:734.

18. Engel H, Z.C. de la Cruz, Jimenez-Abalahin LD, Green WR, et al. Cytopreparatory techniques for eye fluid specimens obtained by vitrectomy. Acta Cytol 1982;26:551–560.

19. Watson P. A slide centrifuge: an apparatus for concentrating cells in suspension onto a microscope slide. J Lab Clin Med 1966;68:494–501.

20. Del Vecchio PR, DeWitt SH, Borelli JI, Ward JB, Wood TA, et al. Application of millipore filtration technique to cytologic material. J Natl Cancer Inst 1959;22:427–431.

21. Chess J, Sebag J, Tolentino Fl, Schepens L, Calderone JP, et al. Pathologic processing of vitrectomy specimens: a comparison of pathologic findings with celloidin bag and cytocentrifugation preparation of 102 vitrectomy specimens. Ophthalmology 1983;90:1560–1564.

22. Barrett DL, King EB. Comparison of cellular recovery rates and morphologic detail obtained using membrane filter and cytocentrifuge techniques. Acta Cytol 1976;20:174–180.

23. Keebler CM. Cytopreparatory techniques. In: Bibbo M, ed. Comprehensive cytopathology. Philadelphia: W.B. Saunders, 1991;881–906.

24. Stulting RD, Leif RC, Clarkson JG, Bobbitt D. Centrifugal cytology of ocular fluids. Arch Ophthalmol 1982; 100:822–825.

25. Barrett DL. Sources of cell loss using membrane filter and cytocentrifuge preparatory techniques. Cytotechnol Bull 1975;12:7–8.

26. Roth J. The colloidal gold marker system for light and electron microscopy. Theory and application. In: Bullock GR, Petrusz VP, eds. London: Techniques in immunocytochemistry. Academic Press, vol. II, 1983.

27. Glauert AM. Fixation, dehydration and embedding of biologic specimens. Amsterdam: Elsevier North Holland Biomedical Press, 1975.

28. Osborn M, Domagala W Immunocytochemistry. In Bibbo M, ed. Comprehensive cytopathology. Philadelphia, W.B. Saunders, 1991;1011–1052.

29. Chess Q, Hajdu SI. The role of immunoperoxidase staining in diagnostic cytology. Acta Cytol 1986;30:1–7.

30. Levitt S, Cheng L, DuPuis MM, Layfield LJ. Fine needle aspiration diagnosis of malignant lymphoma with confirmation by immunoperoxidase staining. Acta Cytol 1985;29:895–902.

31. Zajdela A, Vielh P, Schlienger P, Haye C. Fine-needle cytology of 292 palpable orbital and eyelid tumors. Am J Clin Path 1990;93:100–104.

32. Glasgow BJ, Brown HH, Zaragoza A, Foos RY. Quantitation of tumor seeding in intraocular fine needle aspiration. Am J Ophthalmol 1988;105:538–546.

33. Augsburger JJ, Shields JA, Folberg R, Lang W, O'Hara BJ, et al. Fine needle aspiration biopsy in the diagnosis of intraocular cancer. Cytologic-histologic correlations. Ophthalmology 1985;92:39–49.

34. Glasgow BJ, Straatsma BR, Kreiger A. Analysis of tracts for implantation metastases after intraocular fine needle aspiration of melanomas. Invest Ophthalmol Vis Sci 1991; 32:1197.

CHAPTER

2

Normal Anatomic and Cytologic Features

Accurate cytologic interpretation of ocular specimens requires a fundamental knowledge of normal ocular histology. A general overview is presented here with emphasis on those areas relevant to cytology specimens. There are numerous treatises available for more complete study of ocular anatomy.[1-4]

The average adult eye measures about 25 mm horizontally, 23 mm vertically, and 21 to 26 mm anteroposteriorly (Figures 2–1 and 2–2).[5] The eye has an external approximate volume of 7.6 milliliters (ml), the aqueous has a volume of about 1.5 ml, and the vitreous a volume of 4.0 ml.

The eye is contained in the pear-shaped orbit that has dimensions of about 35 mm vertically, 45 mm horizontally, and 40 to 45 mm anteroposteriorly.[6] The lacrimal gland is located superolaterally in the orbit and is divided by the orbital septum.

CONJUNCTIVA

The conjunctiva covers the posterior surface of the eyelids (palpebral conjunctiva), curves anteriorly at the fornix to reflect onto the anterior surface of the eye as the bulbar conjunctiva (Figures 2–3 and 2–4). There are subtle histologic differences in the conjunctiva of the lid margin, tarsus, fornix, and bulbar conjunctivae.[3] The conjunctiva covering the lid margin and bulbar conjunctiva is a modified nonkeratinized, stratified squamous epithelium. The tarsal and fornix conjunctiva is covered by stratified cuboidal to columnar epithelium of varying thickness (Figure 2–5). This epithelium is unusual because it retains some squamoid features, such as numerous desmosomes, yet has a microvillus surface architecture.[7,8] Goblet cells are abundant over the tarsus, fornix, and specialized areas such as the plica semilunaris. Goblet cells are scarce near the lid margin and adjacent to the cornea at the limbus. Most swabs of the conjunctiva are taken from the inferior fornix and show clusters and single epithelial cells with abundant cytoplasm, eccentric nuclei, and occasional single nucleoli (Figure 2–6). Goblet cells exhibit clear vacuoles filled with mucin. The presence of keratinized epithelium in the conjunctival smear is distinctly abnormal unless the sample is taken from the caruncle or accidentally from the eyelid.

CORNEA

The cornea is covered by five to six layers of a modified stratified squamous epithelium (Figure 2–7).[9] The basal cells are smaller and have a higher nuclear-to-cytoplasmic ratio than the other epithelial cells in the cornea.There are two to three layers of wing cells with interdigitating cytoplasmic processes connected by desmosomes to other wing cells. These attachments may explain why corneal epithelium tends to be removed in sheets. The two top layers are flattened, superficial cells with small, round nuclei and inconspicuous nucleoli. The superficial epithelial cells are normally uniform in size and shape and have many microvilli that form a microplical complex on the external surface of the cornea.[10] Epithelial cells are attached to a basement membrane, beneath which lies Bowman's layer, a specialized layer of collagen. The stroma is composed of lamellar sheets of collagen arranged perpendicularly. The posterior surface of the cornea is covered by Descemet's membrane, and endothelial cells line its posterior surface

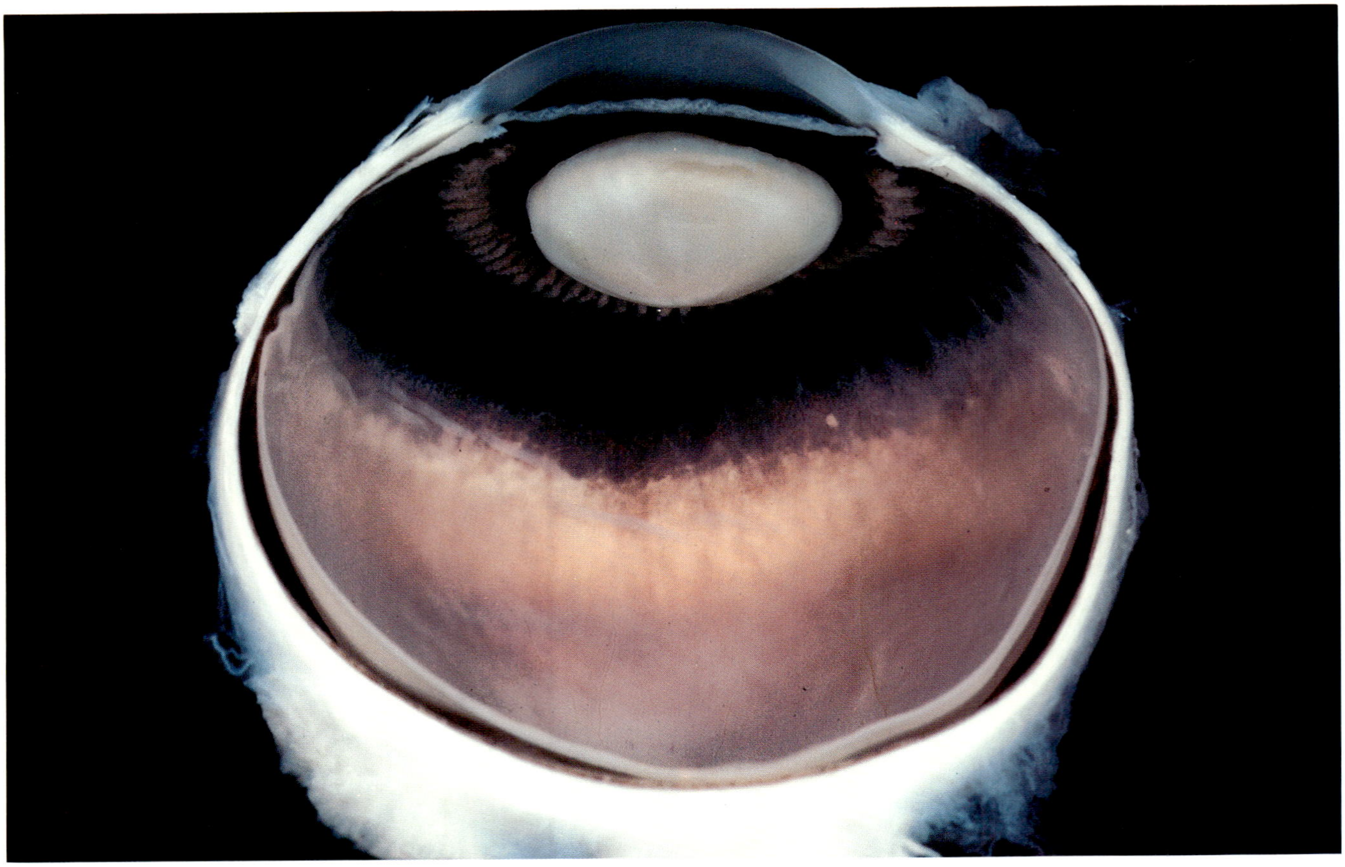

FIGURE 2-1 (ABOVE)

FIGURE 2-2 (BELOW)

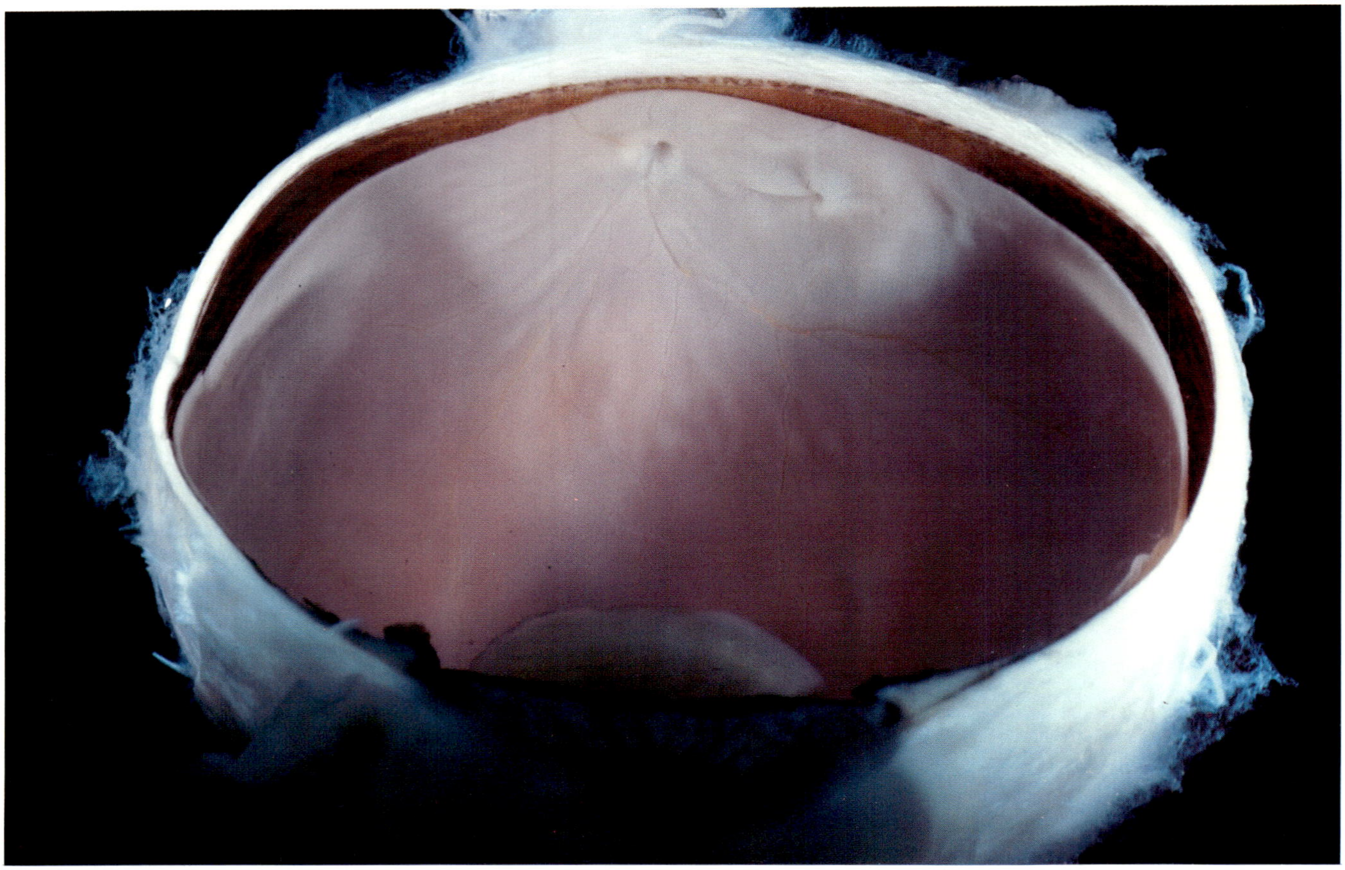

FIGURE 2–1 (ABOVE LEFT) Gross photograph of a sectioned normal human eye. The relationship of the lens, iris, and posterior segment is shown. The ciliary processes (pars plicata) are shown adjacent to the pars plana. The pars plana merges with the retina at the ora serrata.

FIGURE 2–2 (BELOW LEFT) A view of the posterior segment shows the retina, choroid, and sclera.

FIGURE 2–3 Clinical photograph from an everted eyelid demonstrates the normal tarsal conjunctiva and branching vascular pattern.

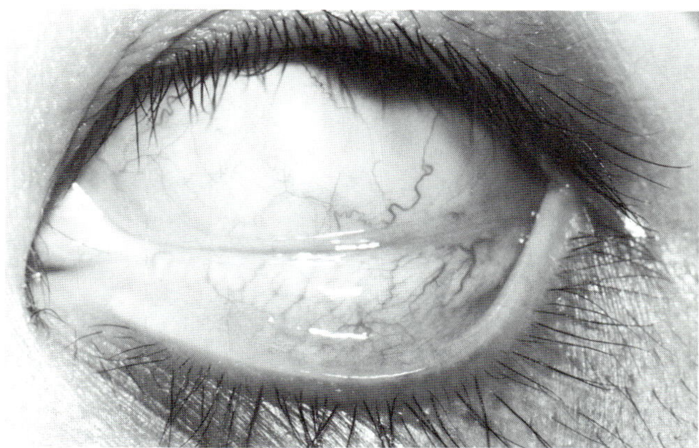

FIGURE 2–4 The inferior fornix becomes visible by simply retracting the lower eyelid.

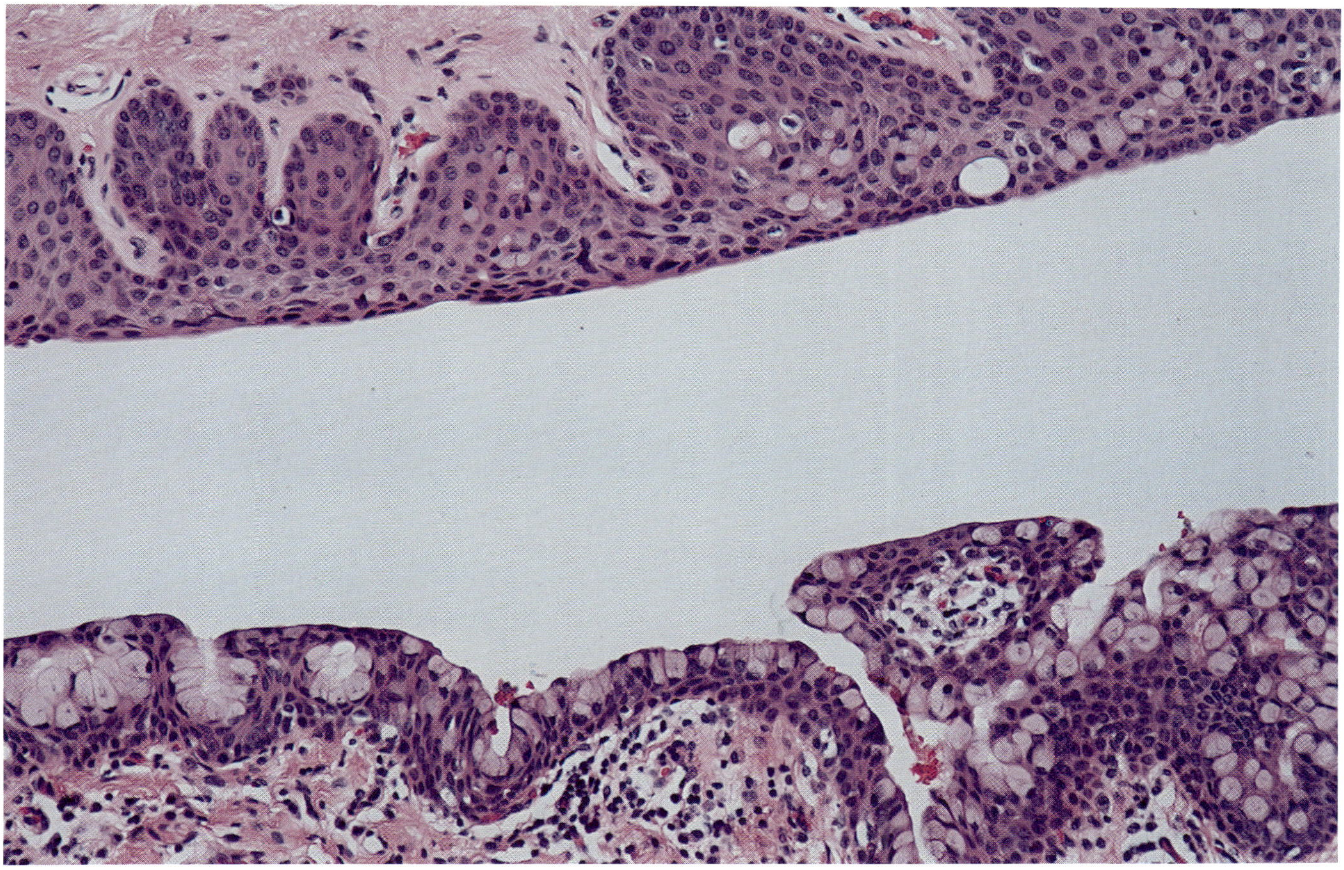

FIGURE 2–5 Photomicrograph shows a section of an eyelid conjunctival surfaces, including tarsus (below) and fornix (above). Note the papillary architecture and numerous goblet cells in the conjunctiva of the fornix. (hematoxylin and eosin, × 45)

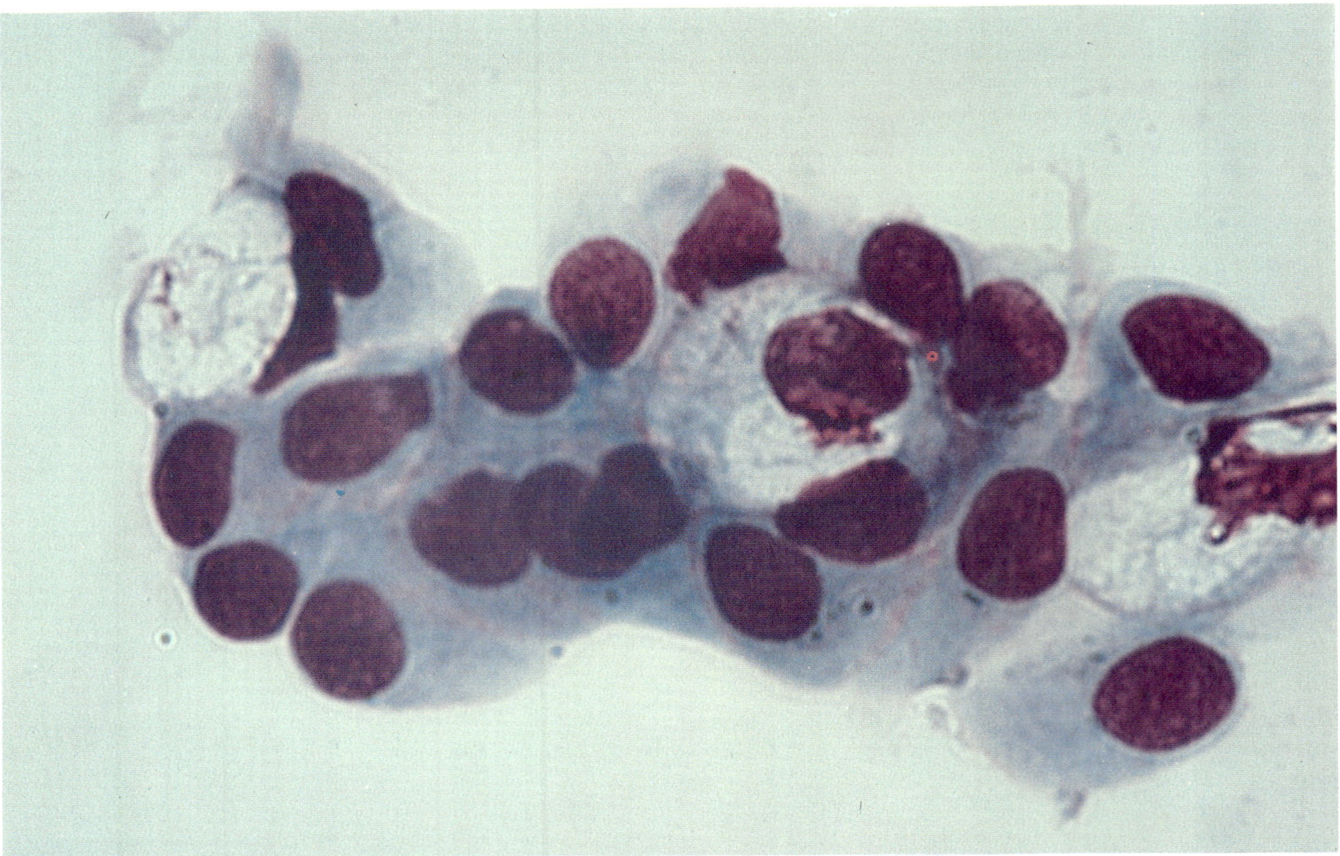

FIGURE 2–6 Photomicrograph shows a cluster of conjunctival epithelial cells and a single goblet cell. (May-Grünwald Giemsa, × 450)

(Figure 2–7). Cytologic surface smears from the normal cornea will demonstrate cohesive sheets of non-keratinized squamous epithelium. Individual cells exhibit intermediate-size, round nuclei with bland and uniform chromatin (Figure 2–8). The presence of keratinized cells in smears from the cornea is abnormal.

IRIS

The iris is the pigmented diaphragm separating the anterior and posterior chambers. It is joined to the ciliary body at the iris root. The anterior surface is composed of a condensed layer of fibroblasts, melanocytes, and collagen fibrils. The iris stroma contains melanocytes, fibroblasts, and blood vessels arranged in a loose network. The posterior surface of the iris is composed of two pigmented epithelial layers (Figure 2–9). The two layers have interwoven microvilli and are attached to each other laterally by desmosomes. The more anterior iris epithelial cells are fusiform in shape and extend myofilamentous cytoplasmic processes into the iris stroma (the dilator muscle). The posterior epithelial cells are columnar in shape. Both layers are densely pigmented.[11]

The iris is rarely sampled by cytologic techniques. However, normal iris may appear in intraocular washings from incidental ocutome cutting of the iris in an attempt to remove vitreous or lens fragments in the anterior chamber (Figure 2--10). Normal iris also may appear in fine needle aspiration specimens of iris neoplasms. In general, normal iris epithelium is so densely pigmented that cellular details are obscured (Figure 2–11). Iris stroma is characterized by the fine reticular meshwork of very cohesive and vascularized stroma.

LENS

The lens is an encapsulated, biconvex structure that is suspended by thin zonules that are attached to the ciliary body. The lens epithelium is located on the internal surface of the capsule. The interior of the lens is composed of cortical and nuclear cells. These hexagonally shaped anucleate cells are joined by interdigitations (Figure 2–12). Because lens cells migrate anteriorly during embryogenesis, the posterior surface of the normal adult lens has no epithelium. Thus, the posterior surface of the lens is covered only by a capsule (Figure 2–13). The lens is generally sampled during vitrectomy or lensectomy. The lens capsule can be identified on cytology prepara-

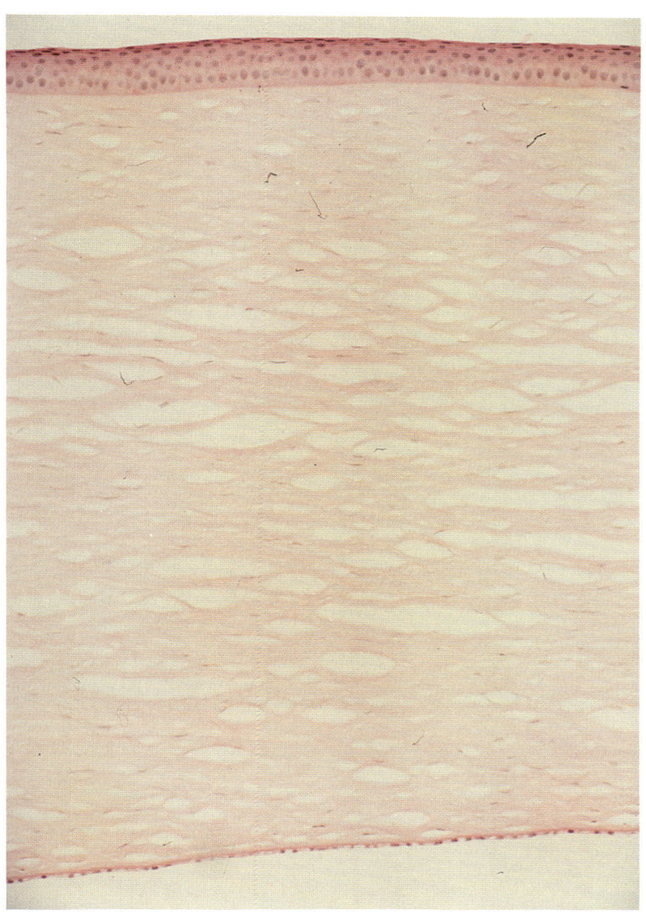

FIGURE 2–7 Photomicrograph of a histologic section of the cornea shows, from top to bottom, epithelium, Bowman's layer, stroma, Descemet's membrane, and endothelium. (hematoxylin and eosin, × 100)

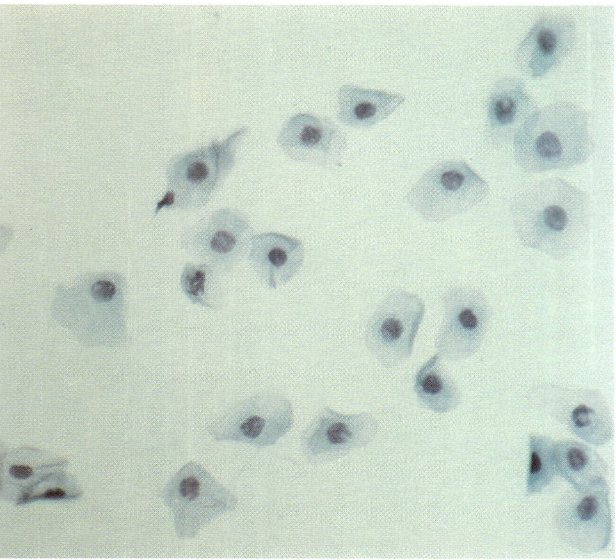

FIGURE 2–8 Smears of corneal epithelium show sheets of uniform nonkeratinized squamous epithelium. The nuclei are round and small. (Papanicolaou, × 250)

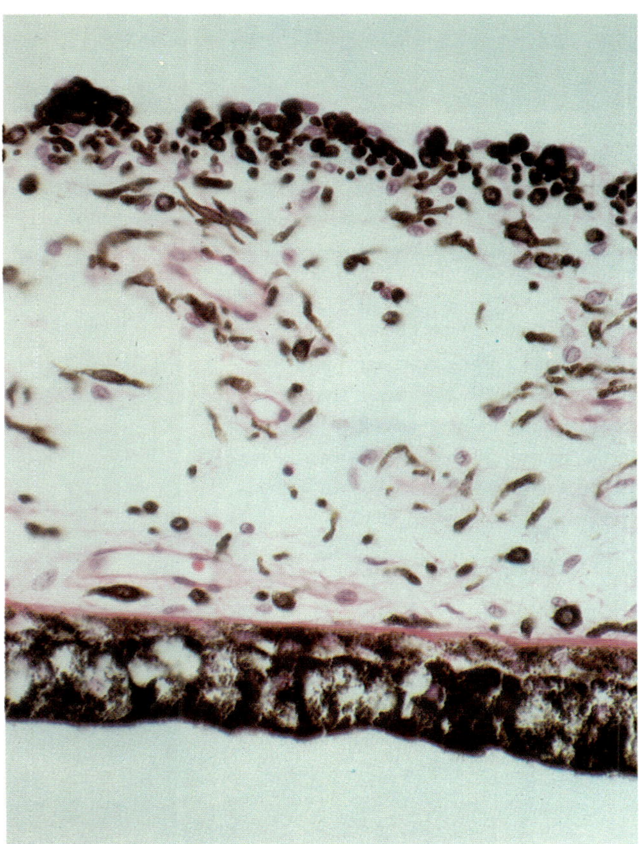

FIGURE 2–9 Photomicrograph of the iris shows a compact, slightly condensed anterior stroma (above), a looser middle stroma, and the posterior pigment epithelial cells. (hematoxylin and eosin, × 125)

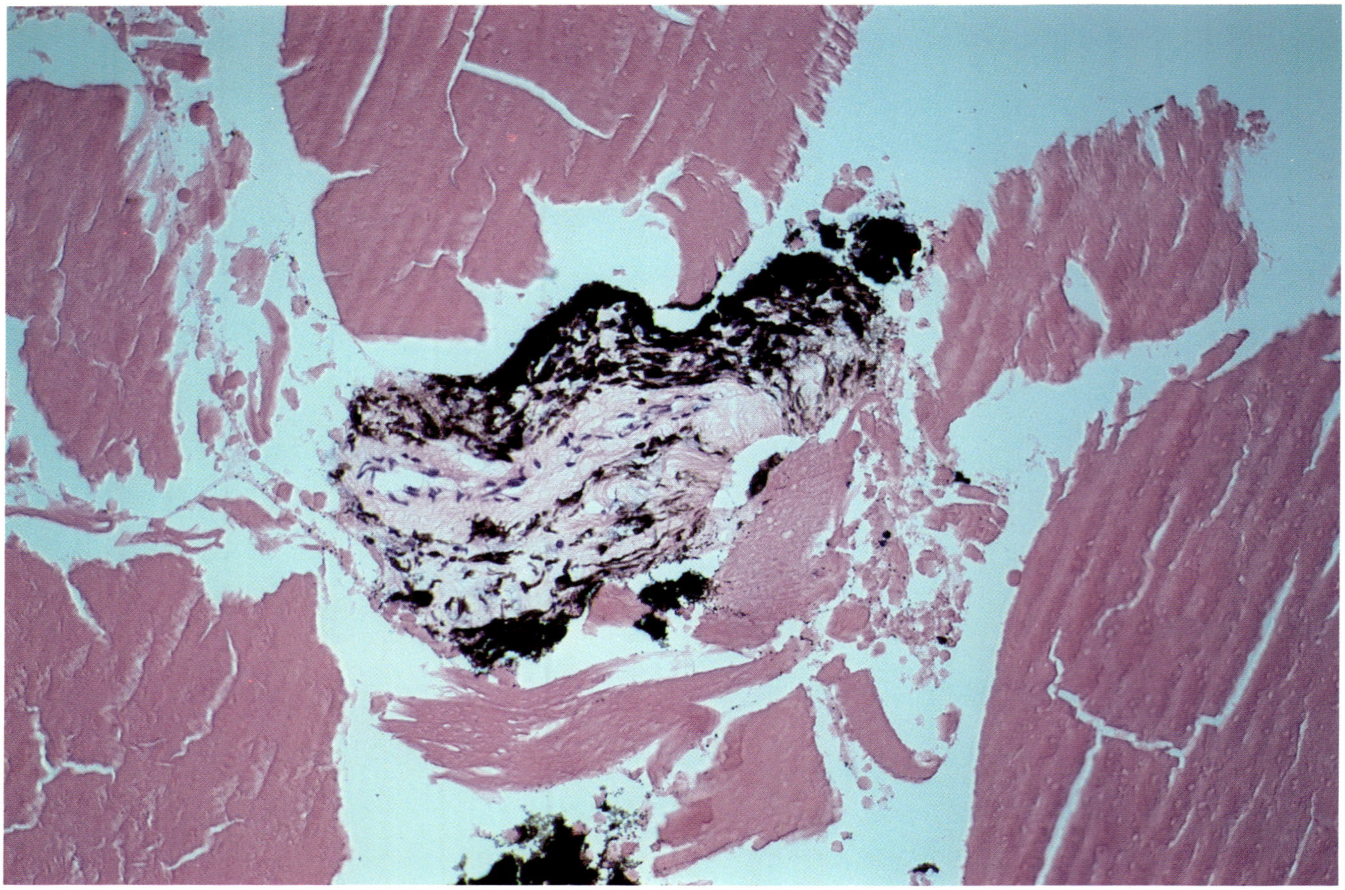

FIGURE 2–10 Photomicrograph of an intraocular washing processed as a cell button displays incidentally removed iris fragments with dense pigment and intervening collagenous stroma. (hematoxylin and eosin, × 50)

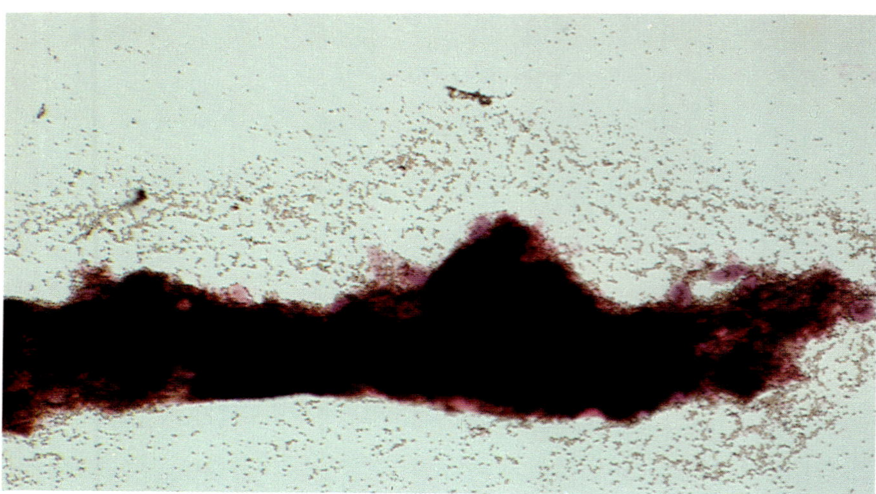

FIGURE 2–11 Photomicrograph of iris from a fine needle aspiration specimen. Dense pigment is seen in cells of stroma and posterior epithelium. A loose collagenous and vascular meshwork is seen. (hematoxylin and eosin, × 125)

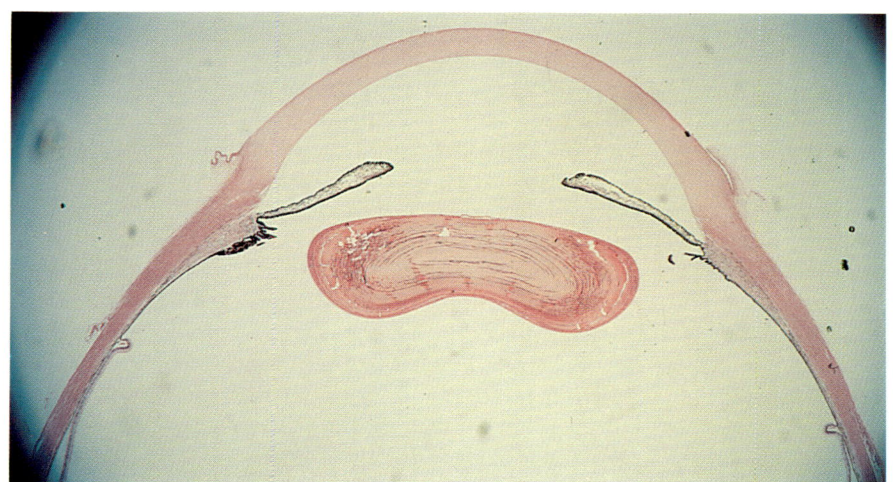

FIGURE 2–12 Section of a normal lens shows eosinophilic lamellar circumferential lens fibers around the central nucleus. (hematoxylin and eosin, × 10)

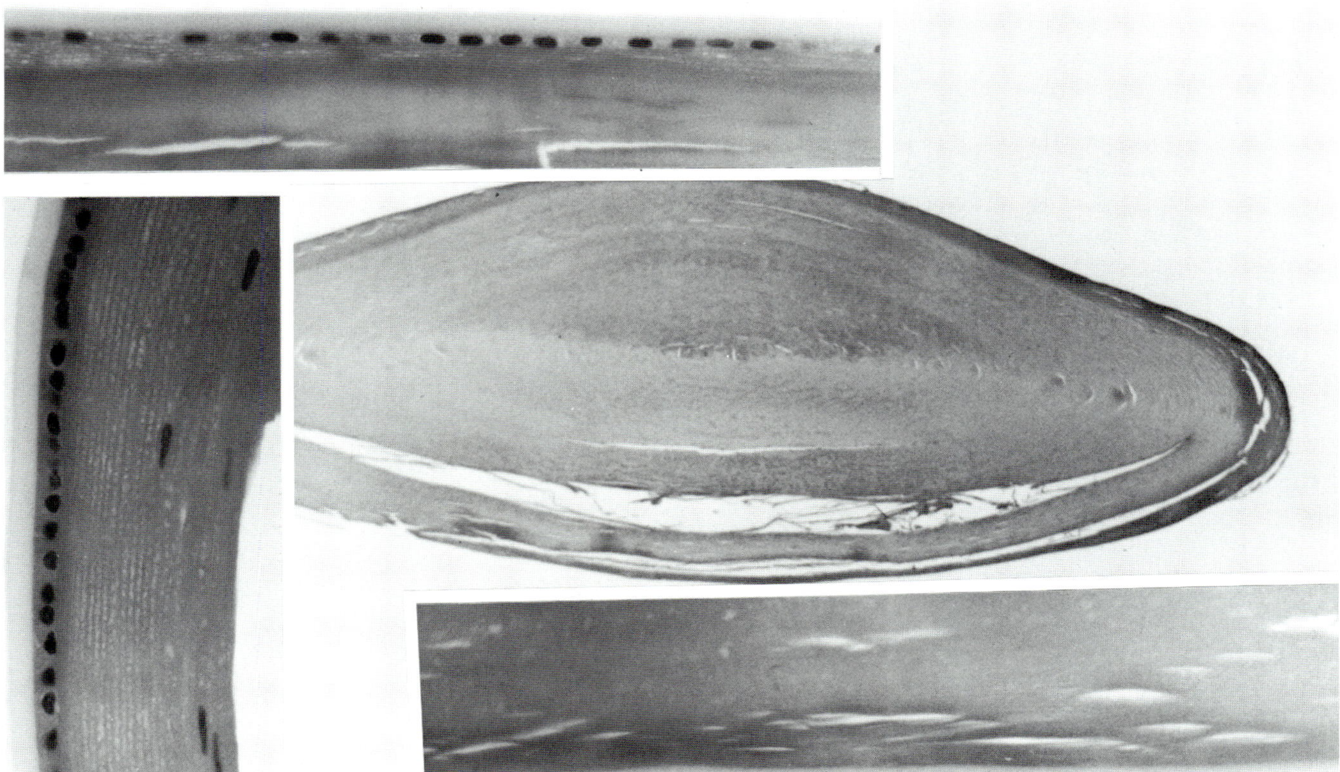

FIGURE 2–13 Higher magnification of different areas of the human lens shows the anterior surface with epithelium (above), the lens bow with nucleated lens fibers (left), and the posterior lens that is devoid of epithelium (bottom).

tions as a translucent (glass) membrane (Figure 2–14). This appears light green with Papanicolaou stain. Cortical fragments taken from the lens periphery or the bow sometimes demonstrate nucleated cells. Lens fragments appear as eosinophilic hexagonal structures by hematoxylin and eosin, and light green with Papanicolaou stain (Figure 2–14).

CILIARY BODY

The ciliary body is composed of the ciliary processes, ciliary muscle, and ciliary epithelium (Figure 2–15). About 70 radially arranged ciliary processes form the pars plicata anteriorly and are joined posteriorly with the smooth portion of the ciliary body, the pars plana (Figure 2–1). The pars plana joins the retina and choroid at the ora serrata. The ciliary body is covered by two layers of epithelium that include an inner nonpigmented layer and an outer pigmented layer (Figure 2–16). Under normal circumstances, ciliary body structures will not appear in vitrectomy specimens. However, ciliary epithelium may be sampled by fine needle aspiration of adjacent tumors. It is important to recognize the two-layered structure of the epithelium with abundant cytoplasm and large pigmented granules (Figure 2–17).[12]

VITREOUS

The vitreous cavity is simply an expanded extracellular space that normally contains 4.0 ml of clear gelatinous substance that is composed largely of water, hyaluronic acid, and collagen.[13] The vitreous normally contains anteroposterior oriented collagen fibrils and occasional macrophages or hyalocytes.[14] The presence of even small numbers of acute or chronic inflammatory cells within the vitreous is distinctly abnormal. The vitreous has distinct attachments to ocular structures.[15] It is attached anteriorly in a circumferential band extending from the posterior pars plana to a few millimeters behind the ora serrata in what has been termed the vitreous base. Traction exerted by the vitreous body at the base results in hyperpigmentation of the underlying pigment epithelium and is evident grossly (Figure 2–1).[16] The vitreous is also attached to the retina over retinal blood vessels and

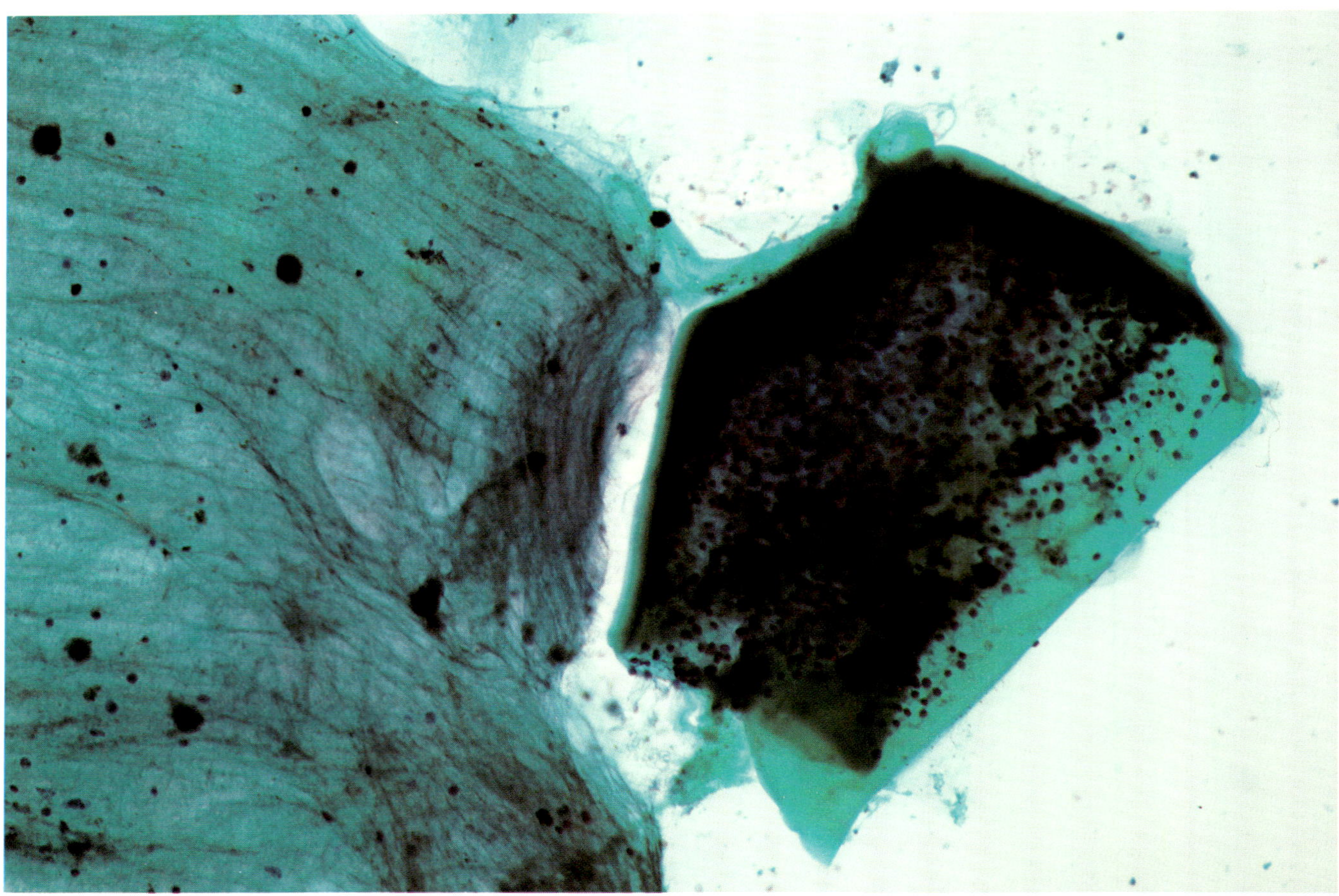

FIGURE 2–14 Cytospin preparations show lens fibers (left) and translucent lens capsule with associated capsular epithelium (right). (Papanicolaou, × 125)

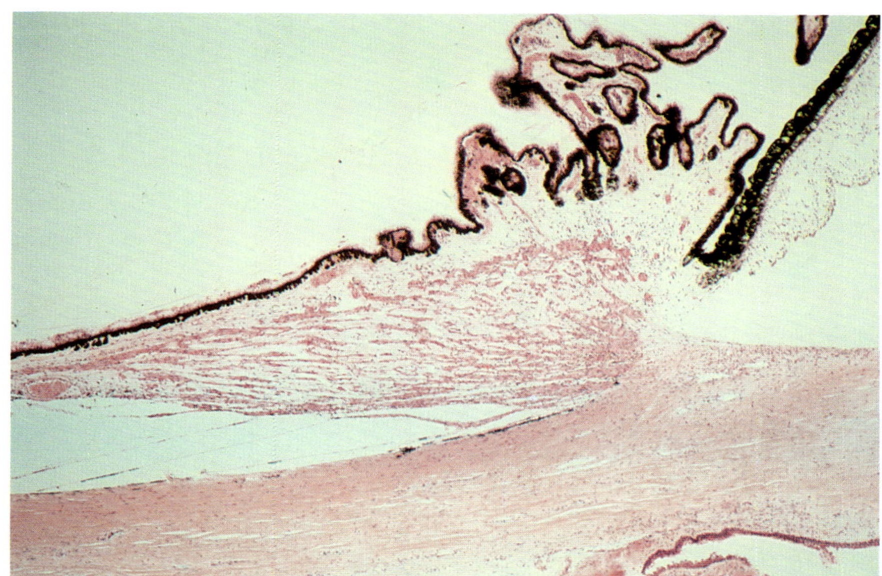

FIGURE 2–15 Sections show the triangular ciliary body that includes ciliary muscle and anterior ciliary processes. (hematoxylin and eosin, × 25)

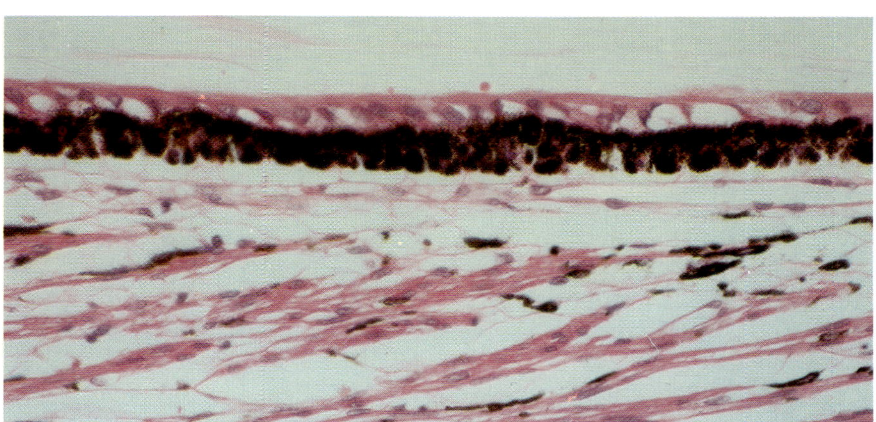

FIGURE 2–16 The pigmented and nonpigmented ciliary epithelium are shown at higher magnification from a section through the pars plana. (hematoxylin and eosin, × 150)

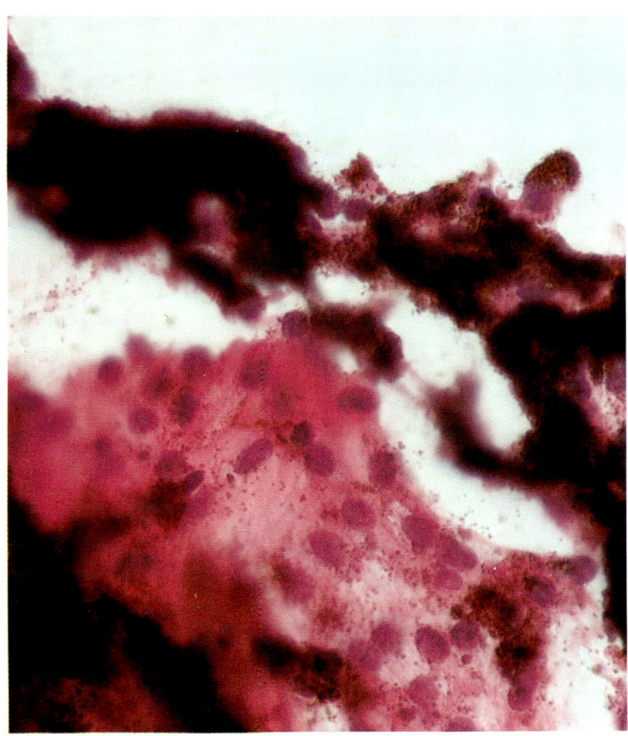

FIGURE 2–17 Smear from fine needle aspirate of normal ciliary body shows closely apposed two layers of pigmented and nonpigmented ciliary epithelium. (hematoxylin and eosin, × 312)

at the optic disc.[17] These attachments are important to understanding vitreous traction, retinal tears, and retinal detachment, for which vitrectomies are sometimes performed.

RETINA

The sensory neuroepithelium of the eye is the retina, which is composed of many layers (Figure 2–18). These include the layer of outer and inner segments of the photoreceptor cells, the outer nuclear layer (cell bodies of photoreceptor cells), the outer plexiform layer, the inner nuclear layer, the inner plexiform layer, the ganglion cell layer, the nerve fiber layer, and the inner limiting lamina (membrane). The retina is loosely attached to the pigment epithelium, which is separated from the choroid by Bruch's membrane. Normal and abnormal retina and pigment epithelium may be sampled in both vitrectomy and fine needle aspiration. In vitrectomy, the retina may be removed as a planned procedure (retinectomy) or it may be removed inadvertently. In cytologic preparations, the retina usually appears as a plexiform pattern of cells with round nuclei and characteristic organoid architecture (Figure 2–19). Often only small fragments of retinal tissue will be present, but can be recognized by the organoid architecture and distinctive nuclear halos (Figure 2–20). Occasionally, ganglion cells may be sampled (Figure 2–19). It is important for the cytologist to report retinal fragments discovered in intraocular washings because full thickness breaks in the retina may lead to retinal detachment. If the surgeon is made aware, the breaks may be closed with cryotherapy, laser, gas injection, or scleral buckle. Fragments of partial-thickness retina that have been stripped in the process of peeling membranes from the retinal surface are not uncommon in intraocular washings and are not regarded presently as clinically significant.

RETINAL PIGMENT EPITHELIUM

The retinal pigment epithelium is a monolayer that lies between photoreceptor outer segments and Bruch's membrane (Figure 2–21). This epithelium has many functions, including matrix production for photoreceptors, phagocytosis of outer segments, barrier protection, and active transport. These cells are large; are polygonal in shape; and contain abundant cytoplasm, round nuclei, and single nucleoli. The cytoplasm contains large distinctive ovoid and elliptical pigment granules (Figure 2–22). The retinal pigment epithelium has a remarkable potential to proliferate and undergo metaplastic transformation.

CHOROID

The choroid underlies the retinal pigment epithelium and is continuous anteriorly at the ora serrata with the ciliary body. It is extremely vascular and contains a layer of capillaries, the choriocapillaris, directly under Bruch's membrane (Figure 2–21). Numerous melanocytes are interspersed with collagen in the choroid. The choroid is not sampled in vitrectomy. However, because many uveal tumors arise from the choroid, it is quite possible to sample choroidal melanocytes in fine needle aspiration. These melanocytes have a stellate shape and contain pigment granules. The cells contain small oval nucleoli (Figure 2–23). They are usually accompanied by the fibrovascular stroma of the choroid. These cells should not be confused with melanoma cells.

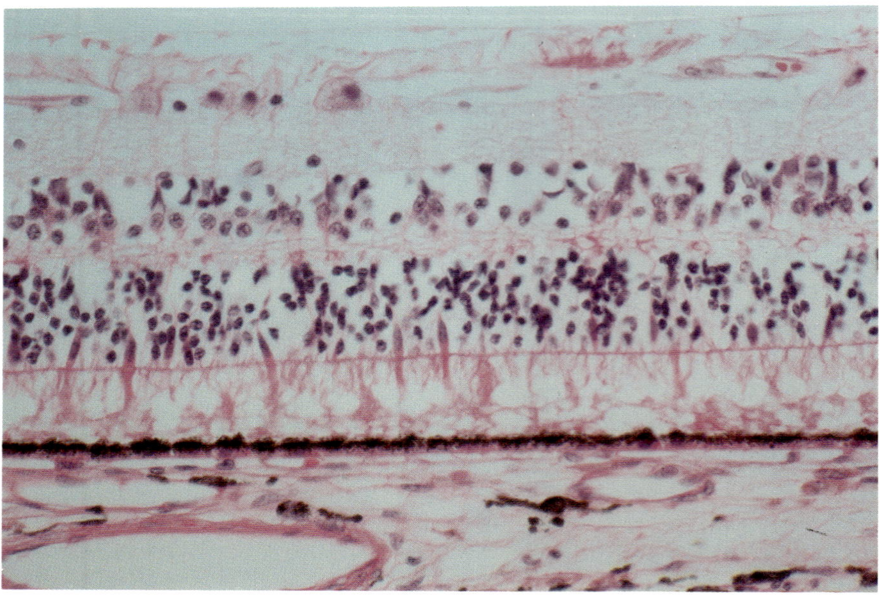

FIGURE 2–18 Section of normal retina shows the characteristic layers. The retinal layers nearest the vitreous cavity are designated inner. (hematoxylin and eosin, × 150)

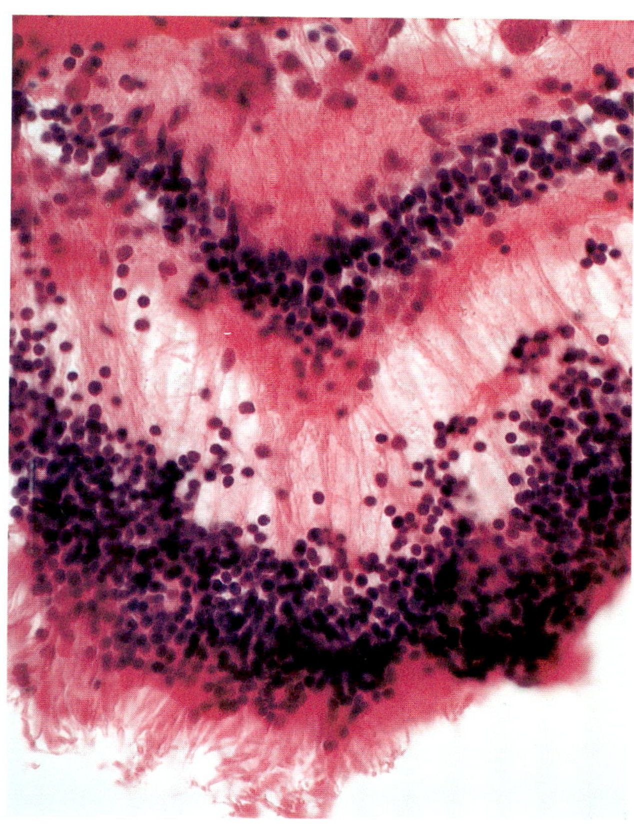

FIGURE 2–19 Cytology preparation dramatically shows normal retinal layers. Rarely will retinal tissue appear so obvious. (hematoxylin and eosin, × 150)

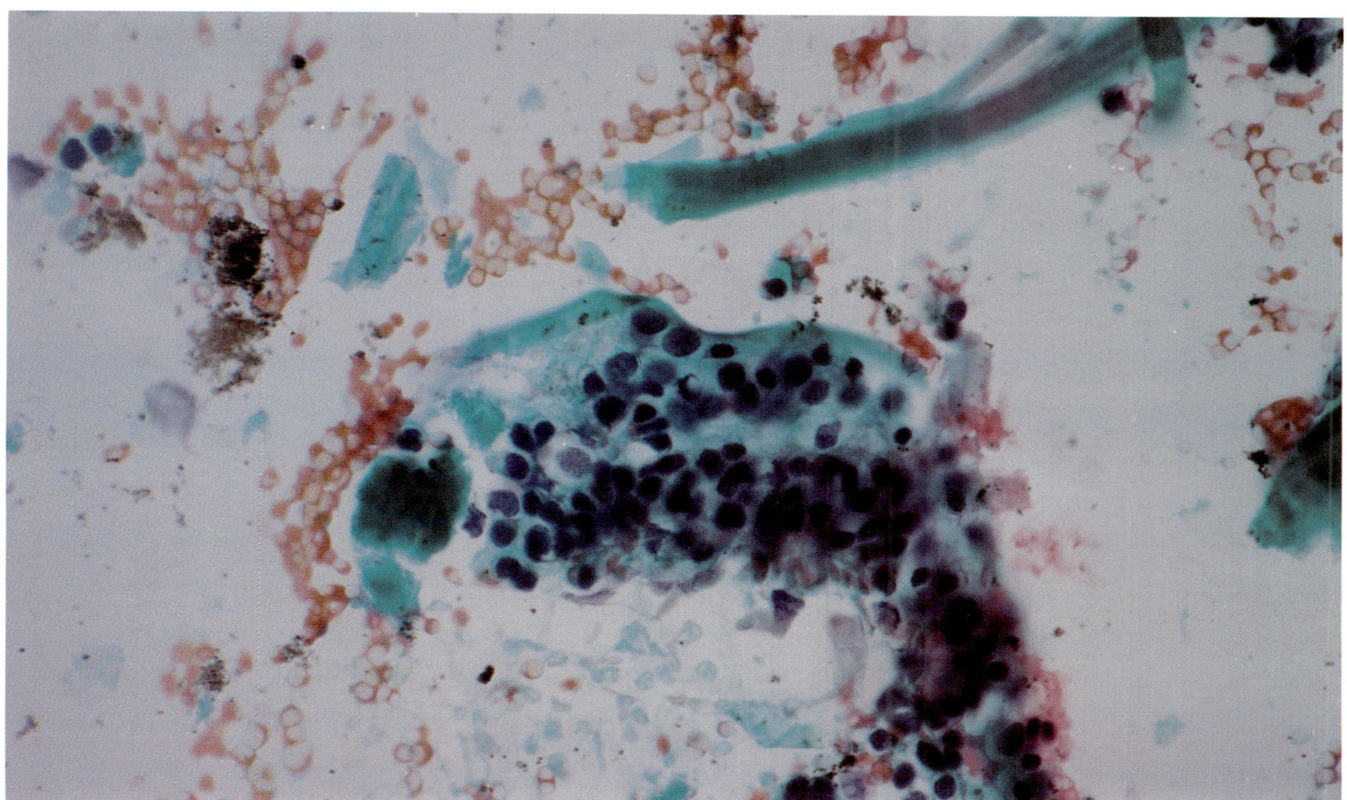

FIGURE 2–20 Cytospin shows small fragment of retinal tissue with an organoid nuclear arrangement and photoreceptor cells. (Papanicolaou, × 150)

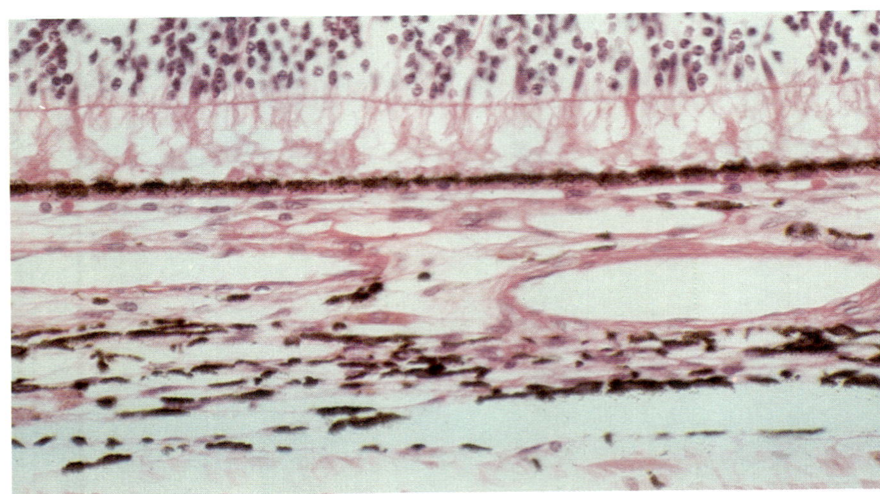

FIGURE 2–21 Section through the posterior segment of a normal eye shows, from top to bottom, retinal pigment epithelium, Bruch's membrane, choriocapillaris, and choroidal vessels. (hematoxylin and eosin, × 150)

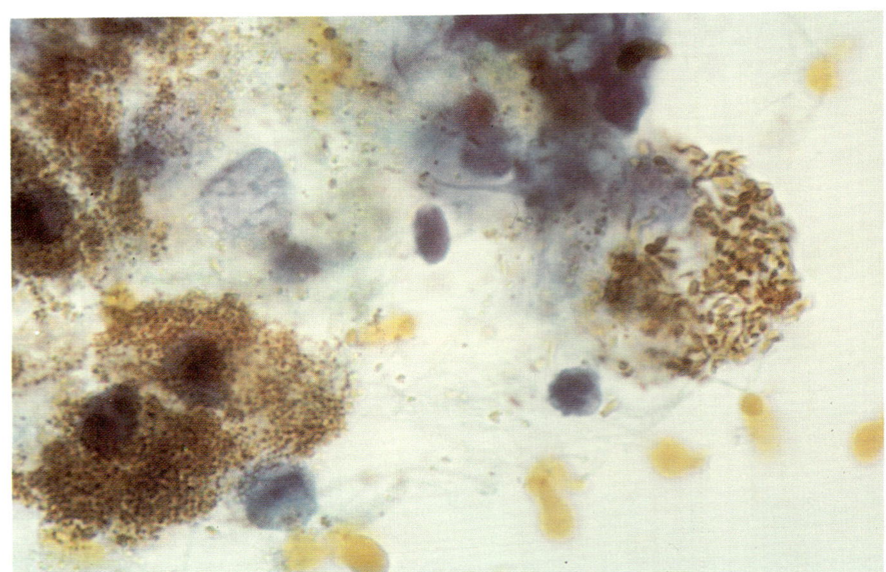

FIGURE 2–22 Cytologic preparation shows cells with round nuclei, abundant cytoplasm, and elliptical granules characteristic of retinal pigment epithelium. (hematoxylin and eosin, × 720)

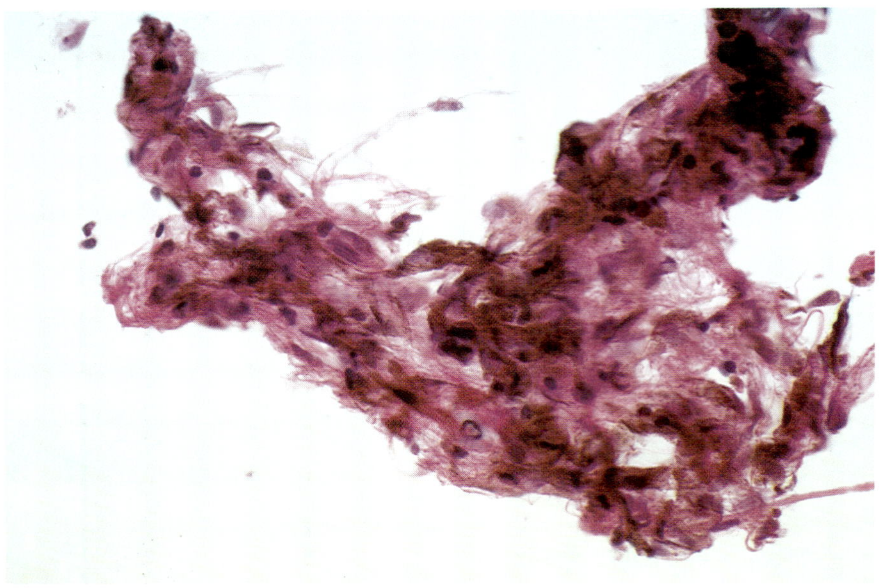

FIGURE 2–23 Cytology smear from the choroid shows intertwined choroidal melanocytes and fibrovascular tissue. (hematoxylin and eosin, × 500)

REFERENCES

1. Jakobiec FA. Ocular anatomy, embryology, and teratology. Philadelphia: Harper & Row, 1982.

2. Hogan MJ, Alvarado JA, Weddell JE. Histology of the human eye. Philadelphia: W.B. Saunders, 1971.

3. Last RJ. Eugene Wolff's anatomy of the eye and orbit. Philadelphia: W.B. Saunders, 1961.

4. Fine BS, Yanoff M. Ocular histology, a text and atlas. New York: Harper and Row, 1972.

5. Stenström S. Untersuchungen über die variation unk kovariation der optishen elemente des menshlickhen auges. Acta Ophthalmol 1946;26:1.

6. Duke-Elder WS. The anatomy of the visual system. In: System of ophthalmology. St. Louis: CV Mosby, 1961;2:410–413.

7. Greiner JV, Covington HI, Allansmith MR. Surface morphology of the human upper tarsal conjunctiva. Am J Ophthalmol 1977;83:892–905.

8. Dark AJ, Durrant TE, McGinty F, Shortland JR, et al. Tarsal conjunctiva of the upper eyelid. Am J Ophthalmol 1974;77:555–564.

9. Hogan MJ, Alvarado JA, Weddell JE. Histology of the human eye. Philadelphia: W.B. Saunders, 1971.

10. Blumcke S, Morgenroth K Jr. The stereo ultrastructure of the external and internal surface of the cornea. J Ultrastruct Res 1967;18:502.

11. Hogan et al., Histology of the human eye, 202–255.

12. Glasgow BJ. Intraocular fine needle aspiration of coronal adenomas. Diagn Cytopathol 1991;7:239–242.

13. Tolentino FI, Schepens CL, Freeman HM. Vitreoretinal disorders, diagnosis and management. Philadelphia: W.B. Saunders, 1976;l–43.

14. Sebag J, Balazs EA. Morphology and ultrastructure of human vitreous fibers. Invest Ophthalmol Vis Sci 1989;30:1867–1871.

15. Foos RY. Vitreoretinal juncture: topographical variations. Invest Ophthalmol Vis Sci 1972;10:801–808.

16. Foos RY. Anatomic and pathologic aspects of the vitreous body. Trans Am Acad Ophthalmol Otolaryngol 1973;77:OP171–OP183.

17. Gartner J. Histologische Beobachtugen uber physiologische vitreovaskulare. Adharenzen Klin Mbl Augen 1962;141:530–545.

18. Foos RY. Vitreous base, retinal tufts, and retinal tears: pathogenic relationships. In: Pruett RC and Regan CCJ, ed. Retina Congress. New York: Appleton-Century-Crofts, 1974.

CHAPTER 3

External Disease of the Conjunctiva, Cornea, and Lacrimal Drainage System

Exfoliative ocular cytology has been used effectively to identify the cell types participating in conjunctival inflammation[1] and to identify specific infectious agents, such as chlamydia.[2] Unfortunately, exfoliative cytology of the cornea and conjunctiva is less reliable for the diagnosis of neoplastic conditions. In this chapter, the utility of cytology in external ocular diseases is addressed with illustrations of selected examples.

ACUTE CONJUNCTIVITIS

Acute bacterial conjunctivitis is one of the most common causes of a red eye. The clinical finding of a red eye with mucopurulent discharge is not specific and may be seen in other conditions, including allergic conjunc-tivitis. While cultures are the best means to determine a specific etiologic agent, conjunctival cytology is a simple way to confirm an acute inflammatory response. However, conjunctival cultures have certain disadvantages. They are expensive. A positive culture may not indicate pathogenicity because bacteria normally inhabit the conjunctiva and eyelid. Because many bacterial conjunctival infections are self-limited, microbial cultures are frequently superfluous. Furthermore, topical antibiotics have a broad spectrum so that initial treatment is usually not based on culture results. In acute bacterial conjunctivitis, cytologic preparations show epithelial cells and numerous segmented polymorphonuclear leukocytes (Figure 3–1). These findings are not specific for bacterial conjunctivitis, but simply indicate a marked acute inflammatory response.

Gonococcal conjunctivitis is noteworthy because it requires immediate systemic treatment. Clinical findings of gonococcal conjunctivitis usually include a red eye and characteristically an intense hyperpurulent discharge. Gram stain reveals gram-negative intracellular diplococci and sheets of neutrophils. Because the disease can be fulminant, initial treatment is predicated on the results of the gram stain.

ALLERGIC CONJUNCTIVITIS

Allergic conjunctivitis includes the clinical subsets of hay fever, atopic keratoconjunctivitis, vernal conjunctivitis, giant papillary conjunctivitis, and contact allergy. All of these disorders produce symptoms of intense itching and tearing. Enlarged conjunctival papillae are seen on clinical examination (Figure 3–2).

Hay fever is the most common form of allergic conjunctivitis.[3] It is seasonal in character, mild in intensity, and shows no corneal involvement.

Atopic keratoconjunctivitis has been associated with atopic dermatitis,[4] occurs in the late teens, shows greater enlargement of papillae in the lower palpebral conjunctiva than in the upper tarsal conjunctiva, and often exhibits eyelid swelling and maceration.[5]

Vernal conjunctivitis is characterized by a seasonal predilection, propensity for elevated large cobblestone papillae of the upper tarsal conjunctiva, and grey limbal nodules called Horner-Trantas dots,[6] that are collections of eosinophils (Figures 3–3 and 3–4).

Giant papillary conjunctivitis is associated with foreign bodies, such as contact lenses, sutures, and prostheses.[7] Giant upper tarsal conjunctival papillae are arbitrarily defined as greater than 0.3 mm in diameter.[8]

Contact allergic conjunctivitis may be produced by ophthalmic medications, including neomycin, bacitracin, atropine, thimerosal, benzalkonium chloride, and chloramphenicol.[9]

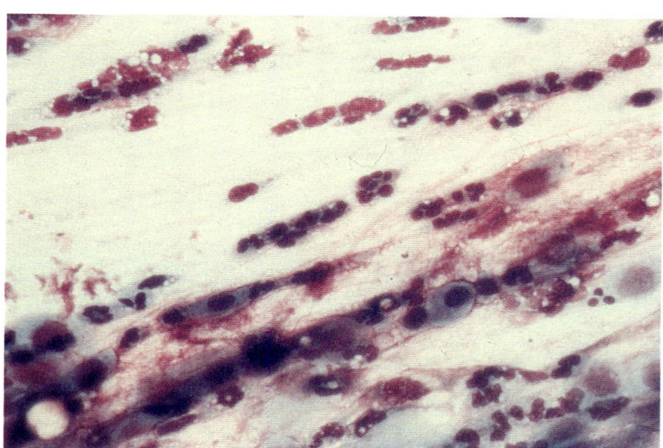

FIGURE 3–1 Giemsa smear shows numerous acute inflammatory cells in a patient with bacterial conjunctivitis (May-Grünwald Giemsa, × 250)

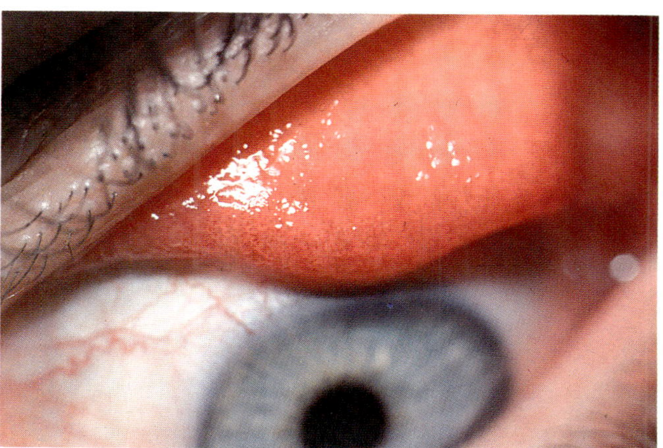

FIGURE 3–2 Clinical photograph shows numerous papillae in a patient with allergic conjunctivitis.

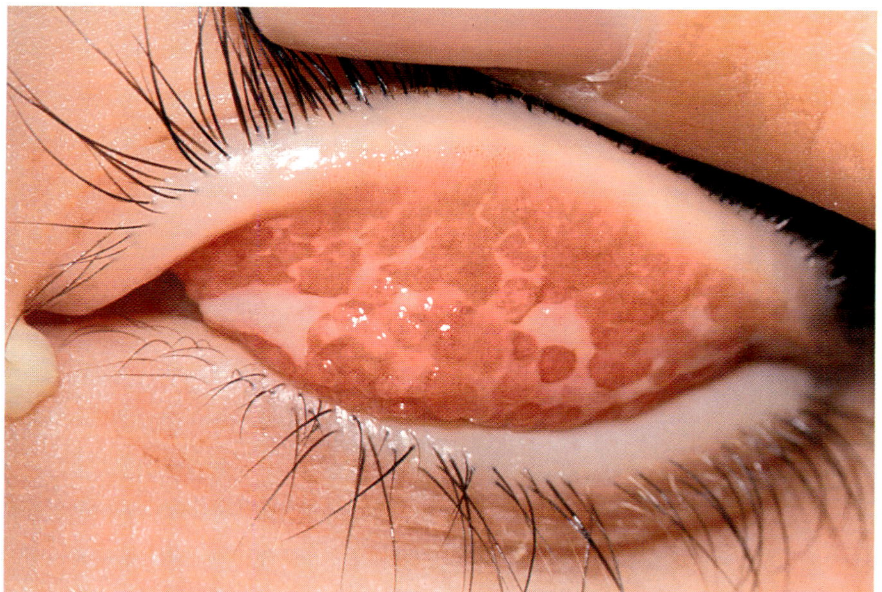

FIGURE 3–3 Clinical photograph shows numerous cobblestone rugae characteristic of vernal conjunctivitis. Photograph provided courtesy of Professor Thomas Petitt, M.D., Jules Stein Eye Institute.

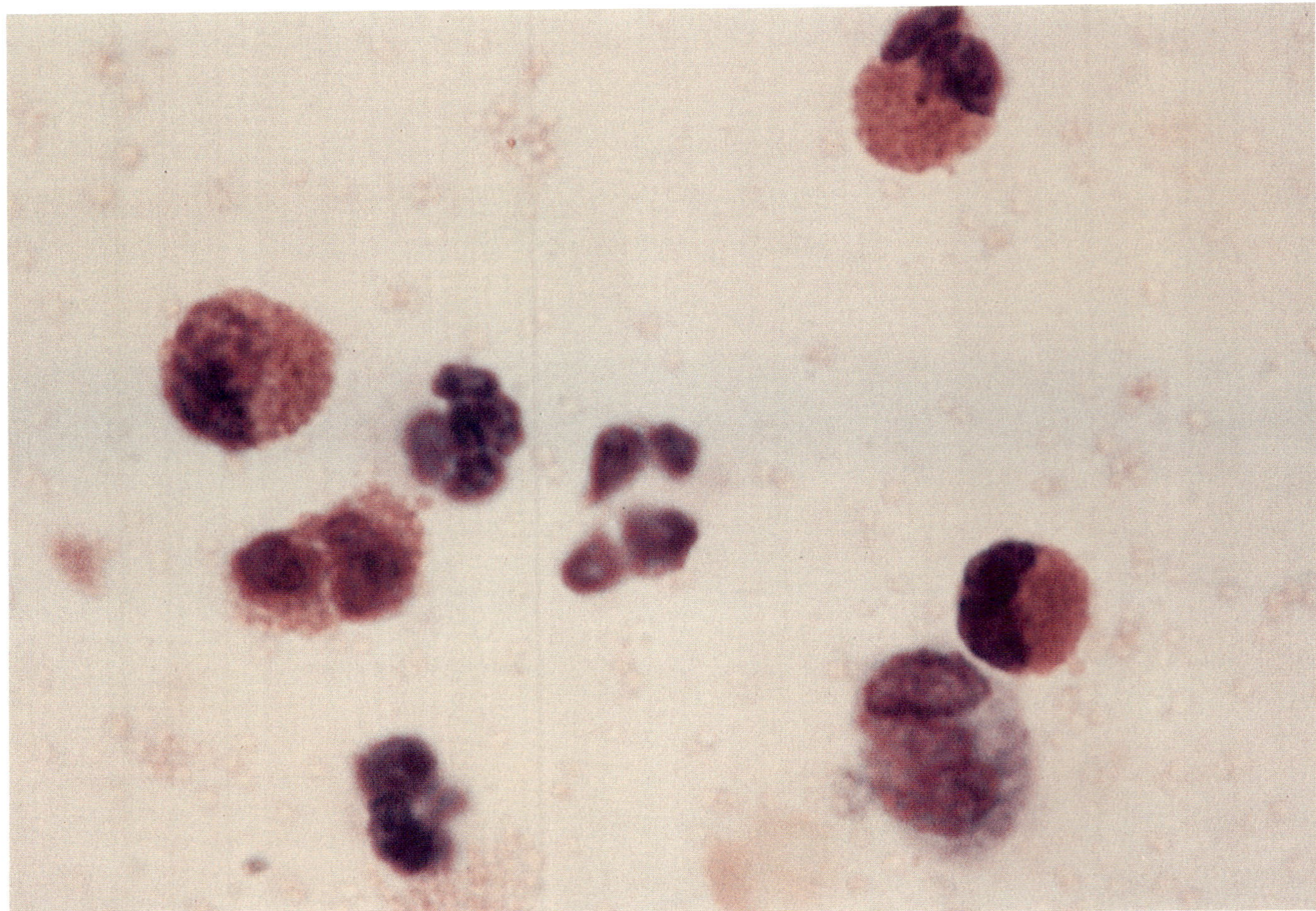

FIGURE 3–4 Photomicrograph of a giemsa preparation shows numerous eosinophils in a patient with allergic conjunctivitis. (Giemsa, × 720). Photograph provided courtesy of Professor Thomas Petitt, M.D., Jules Stein Eye Institute.

Cytologic preparations in all of these allergic disorders may show mast cells, eosinophils, and basophils (Figure 3–4). Eosinophils are not normally present in the conjunctiva; therefore, the presence of any may be significant.[10] However, eosinophils may be absent in allergic conjunctivitis.[11]

ADENOVIRUS

Acute conjunctivitis due to adenovirus infection is characterized clinically by the presence of pale nodules (follicles) in the conjunctiva, a red eye, watery discharge, and preauricular lymphadenopathy (Figure 3–5). Cytologic smears may show numerous lymphocytes of varying sizes, occasional macrophages, and epithelial cell necrosis (Figure 3–6). Occasionally, epithelial cells may exhibit nuclear enlargement, an increased nuclear-to-cytoplasmic ratio and, rarely, intranuclear inclusions. Certain strains of adenovirus, types 3 and 8, may selectively infect conjunctiva.[12]

CHLAMYDIA TRACHOMATIS

Trachoma, caused by the bacterial agent *Chlamydia trachomatis* is a major cause of blindness in the world. In the United States, it manifests as inclusion conjunctivitis characterized by a follicular conjunctivitis (Figure 3–7). Chlamydia is important to diagnose because the disease may require topical and systemic therapy. Historically, the diagnosis has been made by cytology of conjunctival scrapings.[13]

Chlamydial infection begins when the elementary body, a 300 nanometer (nm) particle enters epithelial cells. This particle is enclosed by the host membrane and enlarges to become the initial body (1000 nm). The initial body divides to form numerous elementary bodies. When the cell ruptures, the cycle is continued.

On Giemsa-stained preparations, the initial body is basophilic and helmet shaped (Figure 3–8). The cellular constituents seen on Giemsa-stained preparations from known chlamydia infections include neutrophils (usually

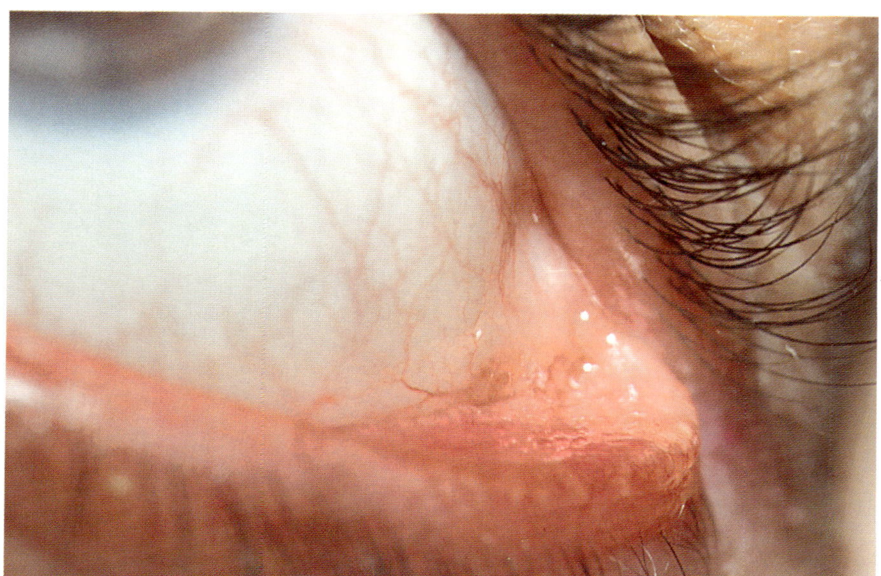

FIGURE 3–5 Clinical photograph of lower tarsus shows large follicles in a patient suspected of a viral conjunctivitis.

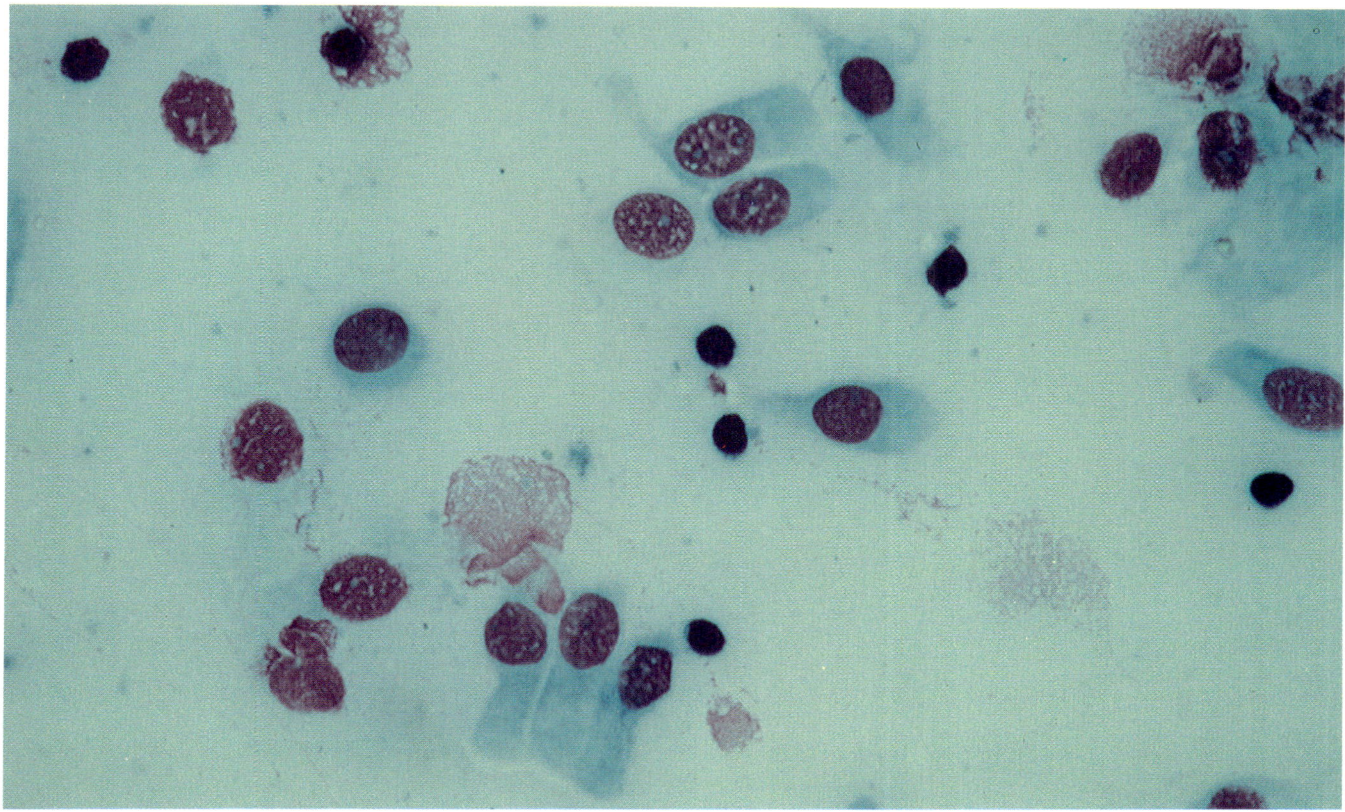

FIGURE 3–6 Photomicrograph of Giemsa smear shows numerous lymphocytes and necrotic epithelium from the patient in Figure 3–5. (May-Grünwald Giemsa, × 250)

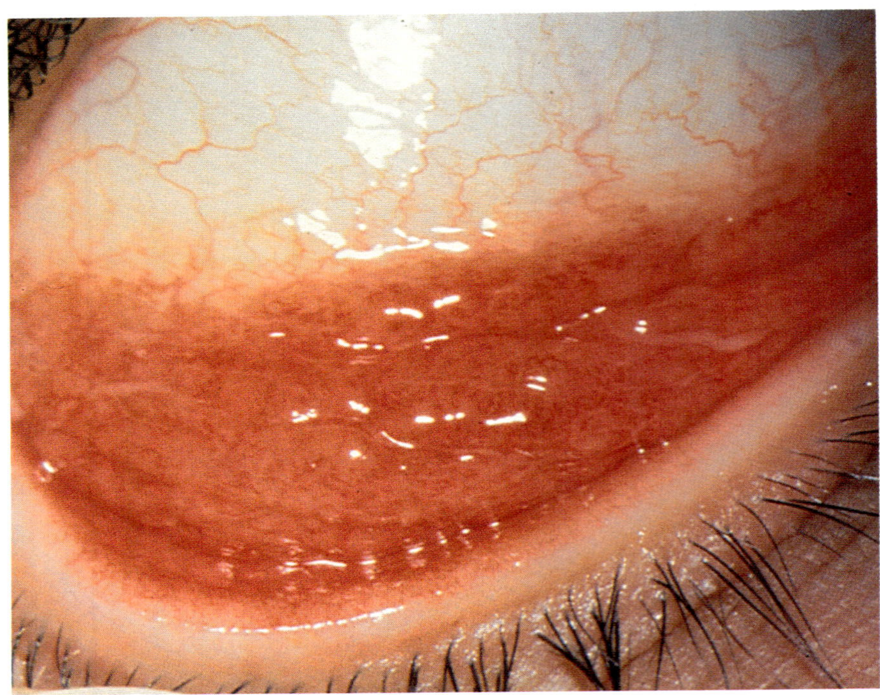

FIGURE 3–7 Clinical photograph shows a follicular conjunctivitis in a patient with documented *chlamydia*. Photograph provided courtesy of Professor Thomas Petitt, M.D., Jules Stein Eye Institute.

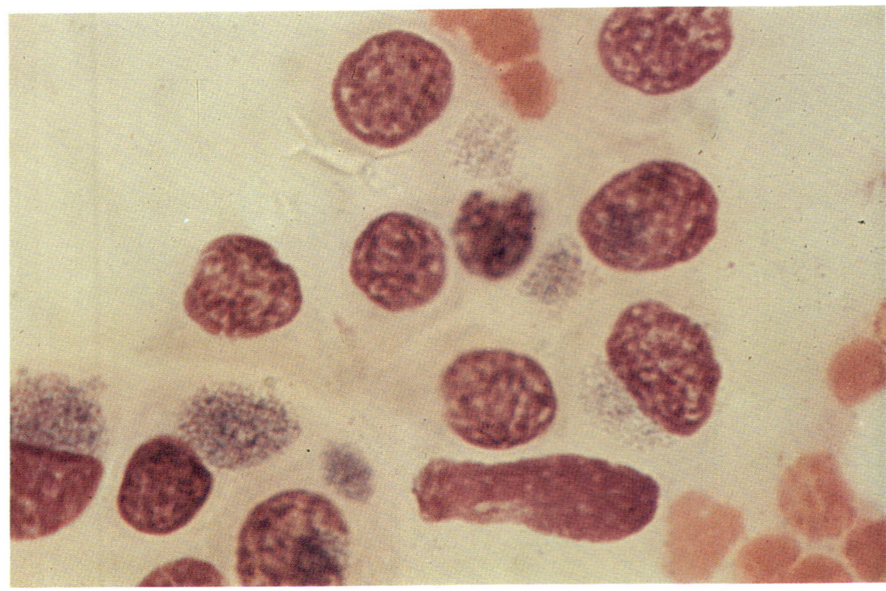

FIGURE 3–8 A Giemsa smear from a patient with culture-proven *chlamydiae* shows numerous intracytoplasmic basophilic bodies. (Giemsa, × 400). Photograph provided courtesy of Professor Thomas Petitt, M.D., Jules Stein Eye Institute.

predominant), lymphocytes, immunoblasts, plasma cells, and occasional multinucleated cells.[14] Because cytoplasmic inclusions are only rarely identified in patients with clinically suspected disease, classic cytology has been largely supplanted by direct immunofluorescence with fluorescein-conjugated monoclonal antibody.[15] However, false-positive results may occur with this test; therefore, a combination of techniques (e.g., smear and culture) has been recommended.[16] In-situ hybridization for chlamydia has been 100% sensitive and specific in animal models.[17] However, in human tissues, the specificity and sensitivity have been reported to be only 80% and 91%, respectively.[18] Papanicolaou staining of smears in gynecologic specimens may be as effective for diagnosis as DNA in-situ hybridization.[19, 20]

CONJUNCTIVAL EPITHELIAL DYSPLASIA

Neoplasia of conjunctival epithelium is classified similarly to cervical neoplasia.[21–25] Severity is graded by the thickness of the conjunctiva, which shows architectural and cytologic abnormalities. The criteria for dysplasia are the loss of normal cellular polarity, nuclear

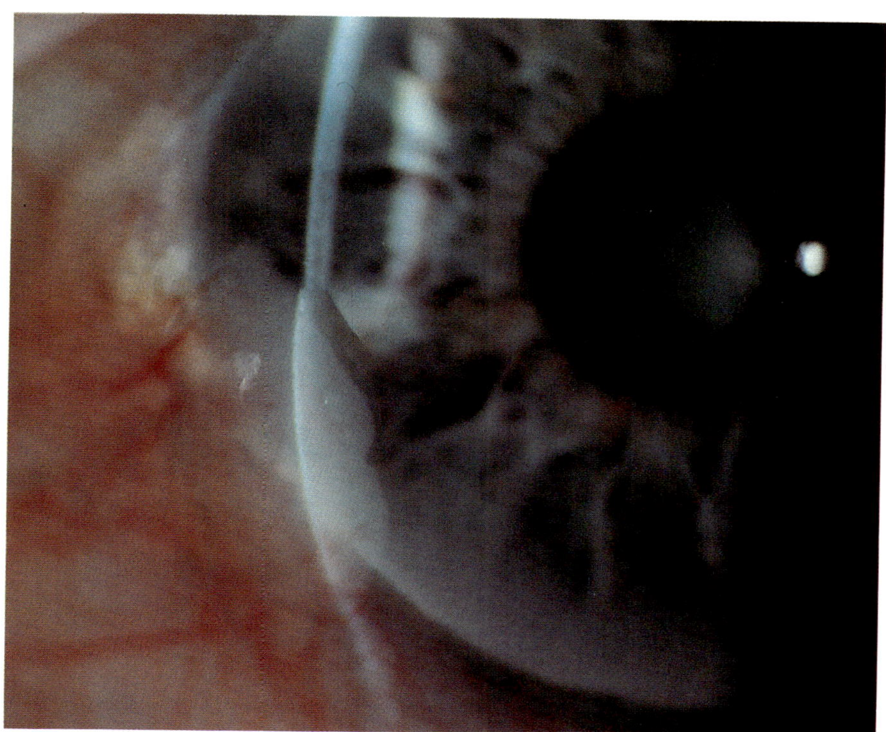

FIGURE 3–9 Clinical photograph shows leukoplakia or a white opalescent-to-opaque covering on the conjunctiva and cornea. Large papillary vessels are present.

enlargement, nuclear irregularity, increased nucleus-to-cytoplasm size ratio, and increased numbers and abnormal forms of mitotic figures. Mild dysplasia involves up to one third of the thickness of the epithelium, moderate dysplasia up to two thirds, and severe dysplasia greater than two thirds. Carcinoma in situ is diagnosed when the entire thickness of the epithelium is involved.

The clinical findings of conjunctival and corneal hyperplasia, dysplasia, and carcinoma in situ are similar. A white epithelium with fimbriated edges and prominent blood vessels produces a surface opacity (Figure 3–9).

Cytologic preparations from conjunctival dysplasia show slightly smaller conjunctival cells with hyperchromatic nuclei, an irregular chromatin pattern, and a high nucleus-to-cytoplasm ratio (Figures 3–10 and 3–11). Although some authors have recommended cytologic examination of the conjunctiva as an excellent screening tool,[26,27] the cytologic grading of conjunctival dysplasia is not as reliable as grading of cervical disease. The reasons for this include the fact that the conjunctiva is very accessible to biopsy and assessment of margins is important in prognostication. Complete excision may be more important in determining recurrence than cell type or degree of dysplasia.[28] In addition, dysplastic lesions of the conjunctiva are frequently keratinizing (Figure 3–12). Cytology from keratinized lesions often yields only surface anucleate squamous cells so that the underlying neoplastic process may escape detection (Figures 3–13 and 3–14).

Papilloma virus, especially type 16, has been iden-

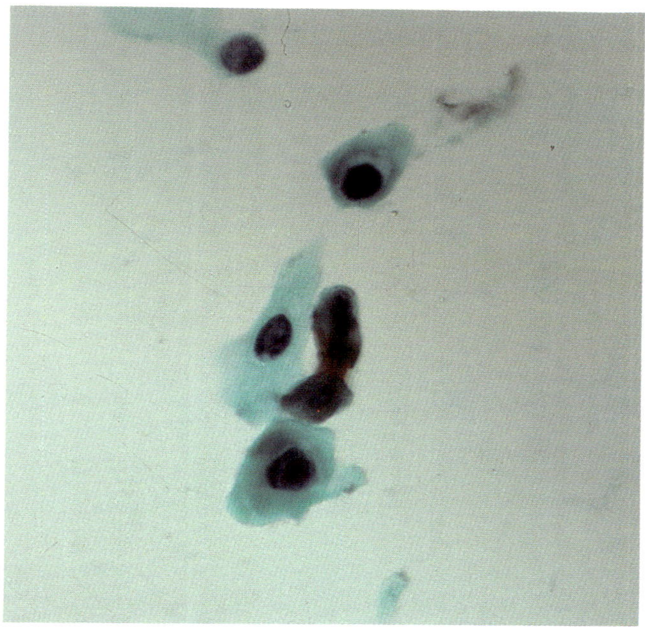

FIGURE 3–10 Alcohol-fixed, Papanicolaou-stained smears show nuclear hyperchromasia and a high nuclear-to-cytoplasmic ratio. (Papanicolaou, × 250)

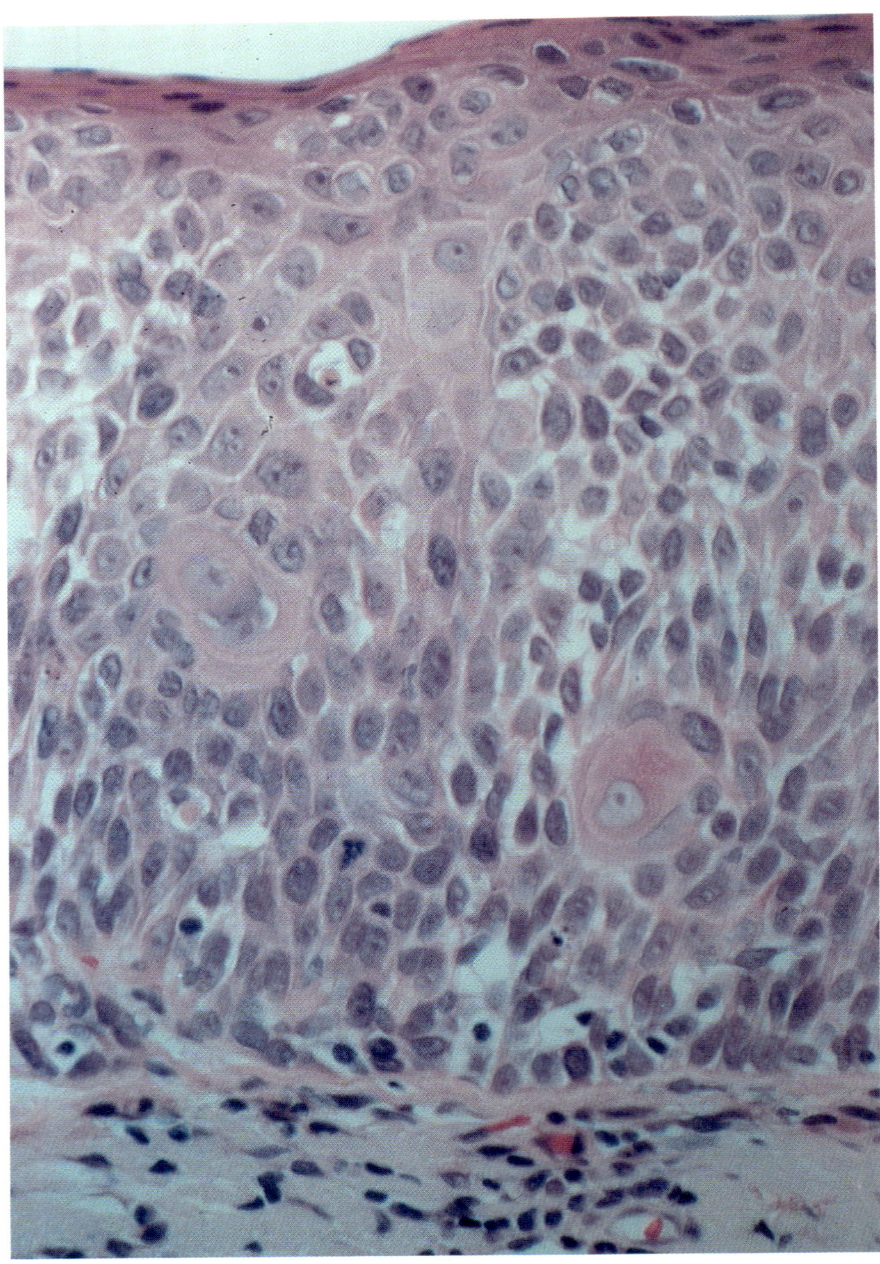

FIGURE 3–11 Biopsy of the lesion shown in Figures 3–9 and 3–10 shows acanthosis, dyskeratosis, abnormal cell polarity, and nuclear irregularities involving greater than two thirds of the thickness of epithelium (severe dysplasia). (hematoxylin and eosin, × 125)

tified in conjunctival swabs of many patients with conjunctival epithelial dysplasias by polymerase chain amplification. The virus is not found in all cases and its significance is yet to be elucidated.[29]

Intraepithelial neoplasia originating in the cornea is extremely rare. Most corneal intraepithelial dysplastic lesions are associated with limbal lesions,[30] and dysplastic lesions confined to the cornea are rare.[31–34] Features considered precancerous in dysplastic lesions include the intracytoplasmic desmosomes, conspicuous tonofibrils, and dyskeratotic bodies and excessive basement membrane.[35,36] Clinically, hyperplastic lesions and dysplastic lesions are often indistinguishable. Accurate cytologic diagnosis of hyperplasia is also difficult.[37]

CANALICULITIS

Inflammation of the lacrimal canaliculus that drains tears from the eye in the nasolacrimal duct is called *canaliculitis*. The most common cause of chronic canaliculitis is *Actinomyces*.[38] *Actinomyces israeli* and *Arachnia propionica* (formerly *Actinomyces propionicus*) have been the major organisms identified.[39–41] The most common clinical findings are swelling of the soft tissue around the puncta and an expressible thick white exudate. Smears of the canalicular discharge show numerous neutrophils and sulfur granules composed of masses of filamentous forms with a fine branching pattern (Figure 3–15). A tissue gram stain (Brown-Brenn) is excellent for

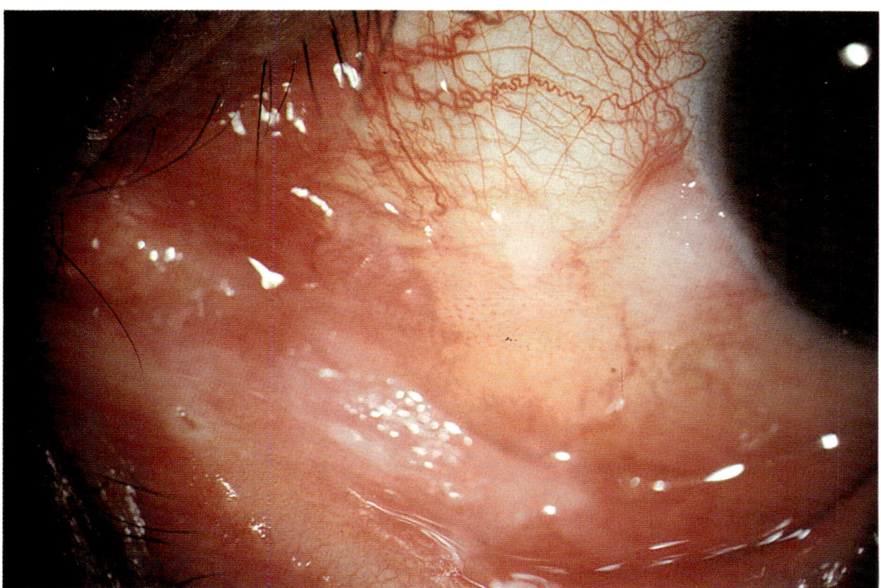

FIGURE 3–12 Clinical photograph shows a very dense white epithelium with prominent vascularity on the conjunctival surface. The process extends to involve inferior fornix and nasal limbus and cornea.

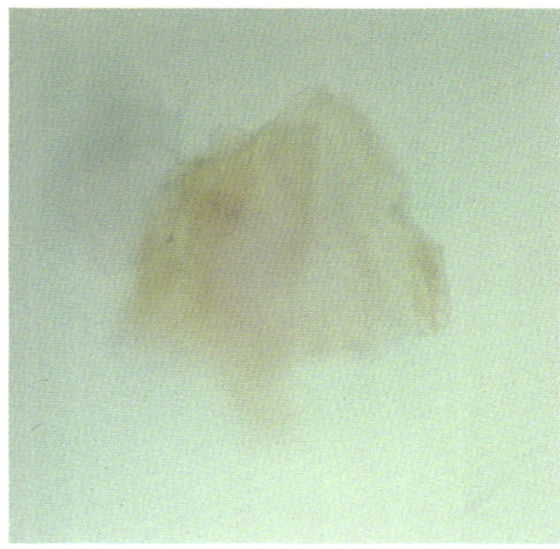

FIGURE 3–13 Photomicrograph of the smear from the patient in Figure 3–12 shows anucleate squamous cells, indicating dense surface keratinization. No clues as to the nature of the underlying dysplastic process are evident. (Papanicolaou, × 720)

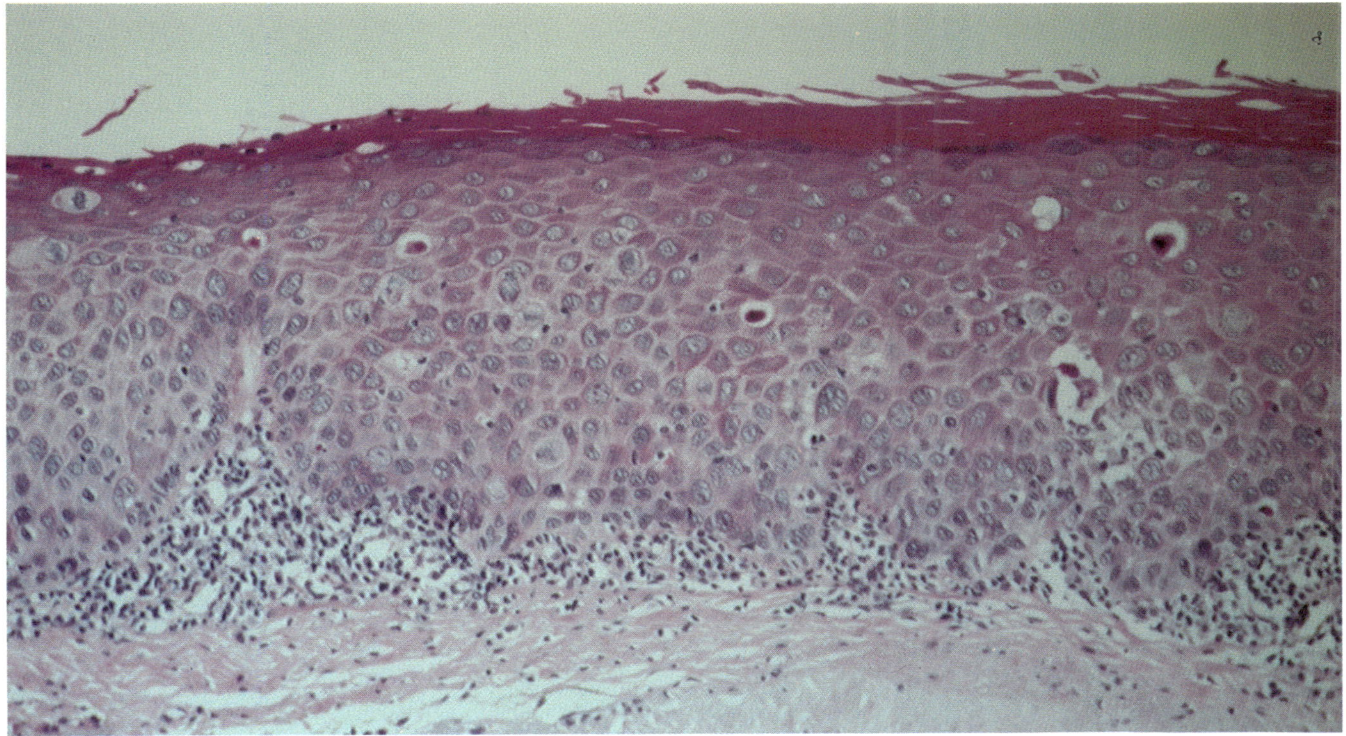

FIGURE 3–14 Sections from the lesion shown in Figure 3–12 show abnormal cell polarity and nuclear atypia involving the entire thickness of epithelium, carcinoma in situ. (hematoxylin and eosin, × 50)

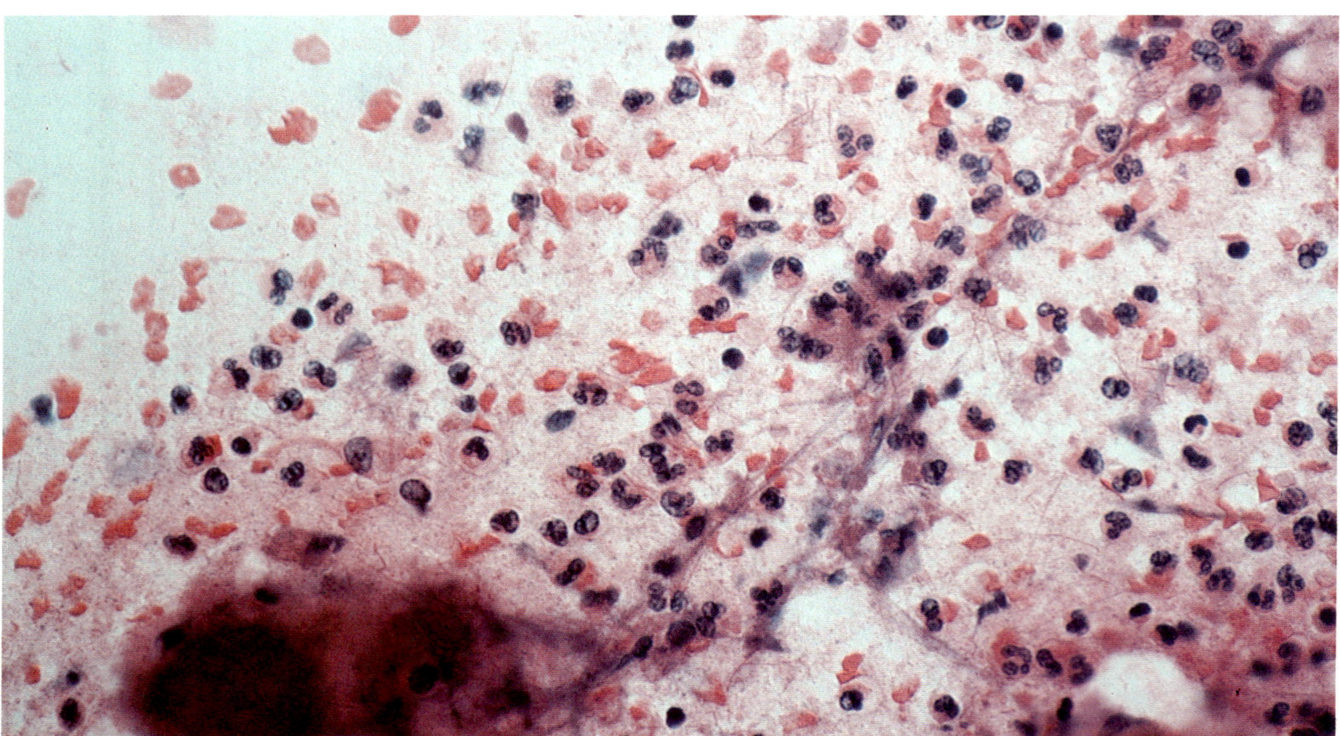

FIGURE 3–15 Smear from the expressed material and curettings from the upper canaliculus of a patient with several days of purulent drainage. Purple staining sulfur granules suggest actinomyces organisms. (hematoxylin and eosin, × 125)

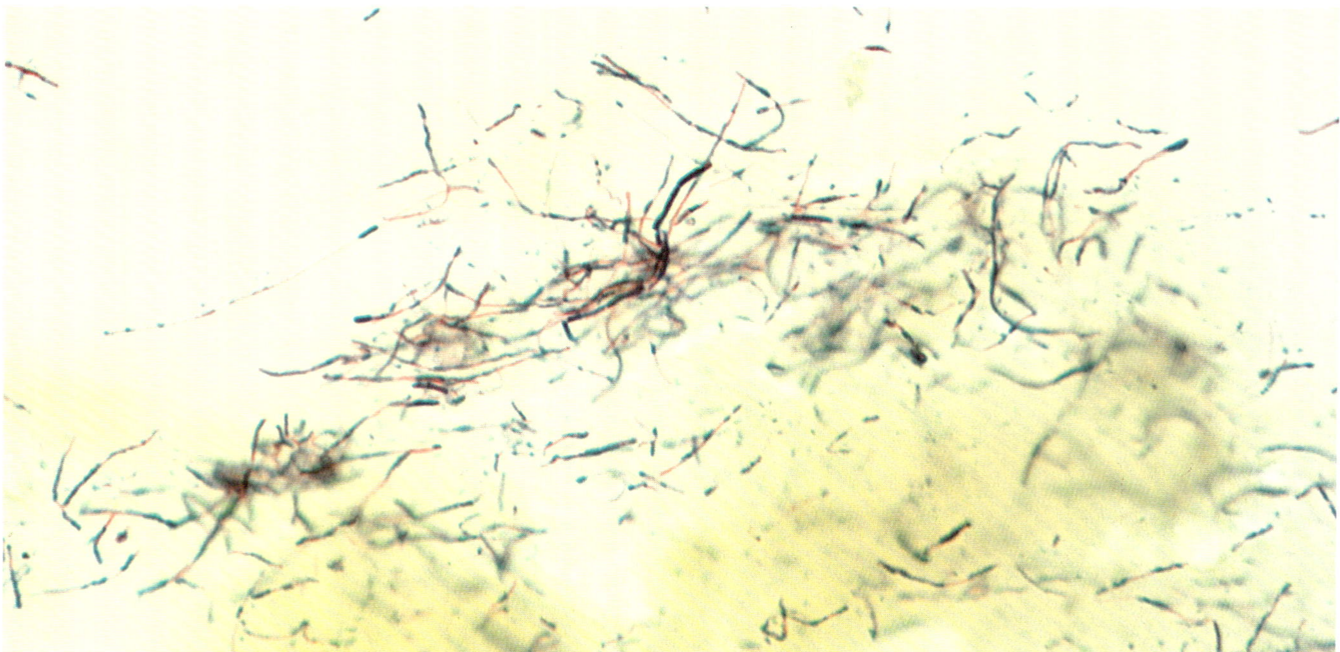

FIGURE 3–16 Brown-Brenn stain of expressed canalicular material from the patient shown in Figure 3–5. *Actinomyces israeli* and *Arachnia propionica* organisms show fine filamental forms with these staining characteristics. (× 250)

demonstrating the actinomyces organism (Figure 3–16). This method is cost effective to diagnose actinomyces canaliculitis.

REFERENCES

1. Herbert H. Pathology and diagnosis of spring catarrh. Brit Med J 1903;2:735.

2. Duggan MA, Pomponi C, Kay D, Robboy SJ. Infantile chlamydial conjunctivitis. A comparison of Papanicolaou, Giemsa and immunoperoxidase staining methods. Acta Cytol 1986;30:341–346.

3. Friedlaender MH, Okumoto M, Kelly J. Diagnosis of allergic conjunctivitis. Arch Ophthalmol 1984;102:1190–1199.

4. Hogan MJ. Atopic keratoconjunctivitis. Am J Ophthalmol 1953;36:937–947.

5. Donshik PC. Allergic conjunctivitis. Int Ophthalmol Clin 1988;28:294–302.

6. Trantas A. Le catarrhe printanier en Turquie. Arch Ophtalmol (Paris) 1905;25:717–731.

7. Richmond PP, Allansmith MR. Giant papillary conjunctivitis. Int Ophthalmol Clin 1981;21:65.

8. Korb DR, Allansmith MR, Greiner JV. Prevalence of conjunctival changes in wearers of hard contact lenses. Am J Ophthalmol 1980;90:336–341.

9. Hatinen A, Terasvirta M, Frakj JE. Contact allergy to components in topical ophthalmologic preparations. Acta Ophthalmol (Copenh) 1985;63:424–426.

10. Butrus SI, Abelson MP. Laboratory evaluation of ocular allergy. Int Ophthalmol Clin 1988;28:324–328.

11. Abelson MB, Madiwale N, Weston JH. Conjunctival eosinophils in allergic ocular disease. Arch Ophthalmol 1983;101:555–556.

12. Thelmo W, Csordas J, Davis P, Marshall KG. The cytology of acute bacterial and follicular conjunctivitis. Acta Cytol 1972;16:172–177.

13. Halberstaedter L, Von Prowazek S. Zur aetiolgie des trachoms. Dtsch Med Wochenschr 1907;33:1285–1287.

14. Wilhemus KR, Robinson NM, Tredici LL, Jones DB. Conjunctival cytology of adult chlamydial conjunctivitis. Arch Ophthalmol 1986;104:693–695.

15. Tam MR, Stamm WE, Handfield HH, Stephens R, Kuo CC, et al. Culture independent diagnosis of *Chlamydia trachomatis* using monoclonal antibodies. N Engl J Med 1984;320,(18):1146–1150.

16. Ugland DN, Jones DB, Wilhemus KR, et al. Diagnosis of adult chlamydial conjunctivitis by use of fluorescein-conjugated monoclonal antibody. Invest Ophthalmol Vis Sci 1985;26 (suppl):272.

17. Horn JE, Kappus EW, Falkow S, Quinn TC. Diagnosis of *Chlamydia trachomatis* in biopsied tissue specimens by using in situ DNA hybridization. J Infect Dis 1988;157:1279–1253.

18. Horn JE, Hammer ML, Falkow S, Quinn TC. Detection of *Chlamydia trachomatis* in tissue culture and cervical scrapings by in situ DNA hybridization. J Infect Dis 1986;153:1155–1159.

19. Gupta PK, Shurbaji MS, Minto JL, Ermatinger SV, Myers J, et al. Cytopathologic detection of *Chlamydia trachomatis* in vaginopancervical (FAST) smears. Diagn Cytopathol 1988;4:224–229.

20. Ghiradini C, Ghinosi P, Raisi O, Rivasi F, Trentini GP. Detection of *Chlamydia trachomatis* in Papanicolaou-stained cervical smears: control study by in situ hybridization. Diagn Cytopathol 1991;7:211–214.

21. Cruess AF, Wasan SM, Willis WE. Corneal epithelial dysplasia and carcinoma in situ. Can J Ophthalmol 1981; 16:171–175.

22. Ash JE, Wilder HC. Epithelial tumors of the limbus. Am J Ophthalmol 1942;25:926–931.

23. Winter FC, Kleh TR. Precancerous epithelioma of the limbus. Arch Ophthalmol 1960;64:208–215.

24. Carrol JM, Kuwabara T. A classification of limbal epitheliomas. Arch Ophthamol 1965;73:545–551.

25. Zimmerman LE. Squamous cell carcinoma and related conditions of the bulbar conjunctiva. In: Boniuk M, ed. Ocular and adnexal tumors, St. Louis: Mosby, 1964:49–74.

26. Dykstra PC, Dykstra BA. The cytologic diagnosis of carcinoma and related lesions of the ocular conjunctiva and cornea. Trans Am Acad Ophthalmol Otolaryngol 1968;73:979.

27. Irvine AR. Epibulbar squamous cell carcinoma and related lesions. Int Ophthalmol Clin 1972;12:72–83.

28. Erie JC, Campbell J, Liesegang TJ. Conjunctival and corneal intraepithelial and invasive neoplasia. Ophthalmology 1986;93:176–183.

29. McDonnell JM, McDonnell PJ, Sun YY. Polymerase chain reaction demonstration of human papillomavirus DNA in tissues and ocular surface swabs of patients with conjunctival epithelial neoplasia. Invest Ophthalmol Vis Sci 1991;32:1285.

30. Waring GO, Roth AM, Ekins MB. Clinical and pathologic description of 17 cases of corneal intraepithelial neoplasia. Am J Ophthalmol 1984;97:547–559.

31. Brown HH, Glasgow BJ, Holland GN, Foos RY. Keratinizing corneal intraepithelial neoplasia. Cornea 1989;8: 220–224.

32. Robertson MC. Corneal epithelial dysplasia. Ann Ophthalmol 1984;16:1147–1150.

33. Geggel HS, Friend J, Boruchoff A. Corneal epithelial dysplasia. Ann Ophthalmol 1985;17:27–31.

34. Cameron JA, Hidayat AA. Squamous cell carcinoma of the cornea. Am J Ophthalmol 1991; 111:571–574.

35. Dark AJ, Streeten BW. Preinvasive carcinoma of the cornea and conjunctiva. Br J Ophthalmol 1980;64:506–514.

36. Campbell RJ, Bourne WM. Unilateral central corneal epithelial dysplasia. Ophthalmology 1981;88:1231–1283.

37. Jakobeic FA, To KW. Corneal epithelial dysmaturation. Invest Ophthalmol Vis Sci 1991;32:1019.

38. Pine L, Hardin H, Turner L, et al. Actinomycotic lacrimal canaliculitis. Am J Ophthalmol 1960;49:1278–1288.

39. Pine L, Georg LK. Reclassification of *Actinomyces propionicus*. Int J Syst Bacteriol 1969;19:267–272.

40. Buchanan BB, Pine L. Characterization of a propionic acid producing actinomycete. *Actinomyces propionicus sp. nov.* J Gen Microbiol 1962;28:305–323.

41. Seal DV, McGill J, Flanagan D, et al. Lacrimal canaliculitis due to *Arachnia (Actinomyces) propionica.* Br J Ophthalmol 1981;65:10–13.

CHAPTER

4 Abnormalities of the Lens

The human lens has a limited response to a wide range of noxious stimuli. The common final end point is cataract, a lenticular opacity produced by degenerative changes in crystalline proteins. Cataracts may form in different parts of the lens depending on the stimulus. Although often distinctive clinically, they are not distinguishable cytologically because the cataractous lens is emulsified by ultrasound (phacoemulsification) or chopped by a vitrectomy cutting instrument. In this chapter, those lens abnormalities that can be diagnosed by cytologic examination are illustrated.

CONGENITAL CATARACTS

There are many clinical types of congenital cataracts, including anterior polar, posterior polar, zonular, sutural, membranous, and filiform. The clinical classification is based on location and slit lamp appearance. These types of cataracts cannot be distinguished in most cytologic preparations because they are generally removed by emulsifying the lens with a cutting instrument (e.g., ocutome). Despite the distortion in anatomic organization, numerous nuclei are often apparent in lens fibers of congenital cataracts obtained by emulsification (Figure 4–1). Rubella produces numerous changes in the eyes of infants, including retinopathy, nongranulomatous uveitis, glaucoma, and congenital cataract (Figure 4–2).[1] Rubella-induced cataracts have been reported to show characteristic retention of nuclei in lens fibers (Figures 4–3 and 4–4).[2–6] This feature can be seen in other types of congenital cataract.[7,8]

ADULT CATARACTS

Most adult cataracts have opacities in multiple locations within the lens. Commonly, they are opaque in both the centrally located nucleus (nuclear sclerosis or nuclear cataract)(Figure 4–5) and the peripherally located cortex (cortical cataract). Less often, opacities are present immediately beneath the capsule (subcapsular cataract). Some subcapsular cataracts are associated with trauma. Histologically, nuclear sclerotic cataracts show degenerative melding of lenticular fibers (Figure 4–6). Severely degenerated cortical lens fibers manifest as round eosinophilic globules (globular or Morgagnian degeneration). In cytologic preparations, Morgagnian globules can be identified (Figure 4–7). Sections of anterior and posterior subcapsular cataracts show a plaque of fibrous tissue and, presumably, lens epithelial cell proliferation immediately beneath the capsule (Figure 4–8).[9] These cataracts are not usually sampled by extracapsular cataract extraction techniques.

Longstanding cataracts may become calcified [10] (cataracta calcarea) presumably by a dystrophic process.[11] In cytologic specimens, calcification appears as basophilic crystals, variable in size and shape, that exhibit birefringence (Figures 4–9 and 4–10). These crystals have been identified as calcium oxylate by x-ray diffraction and electron diffraction.

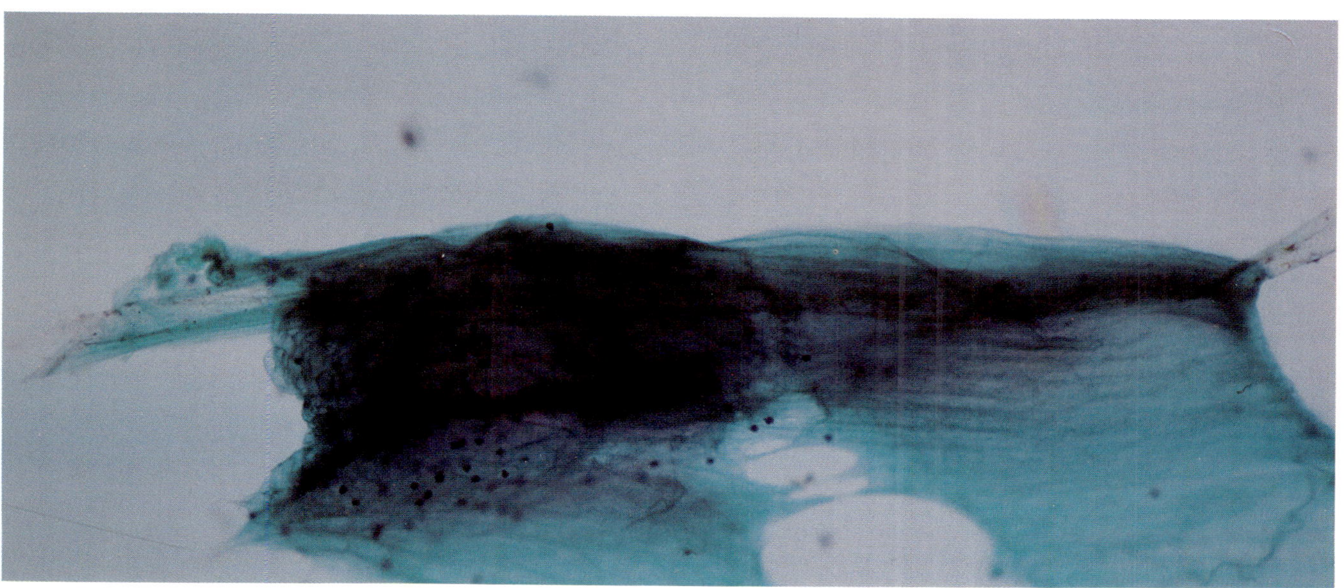

FIGURE 4–1 Intraocular washing from a patient with a congenital cataract. Nuclei are seen within many lens fibers. (Papanicolaou, × 125)

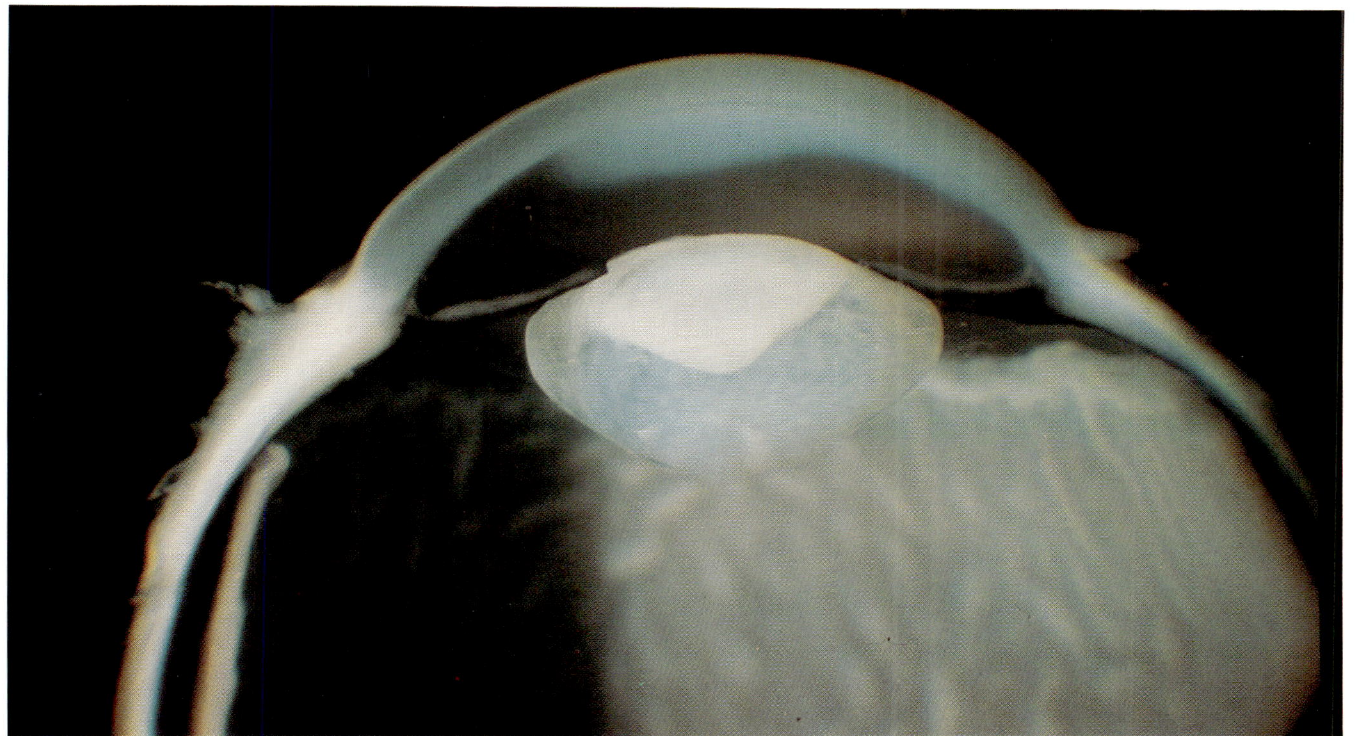

FIGURE 4–2 Gross photograph shows a plaque opacifying the lens (cataract) from a patient with congenital rubella.

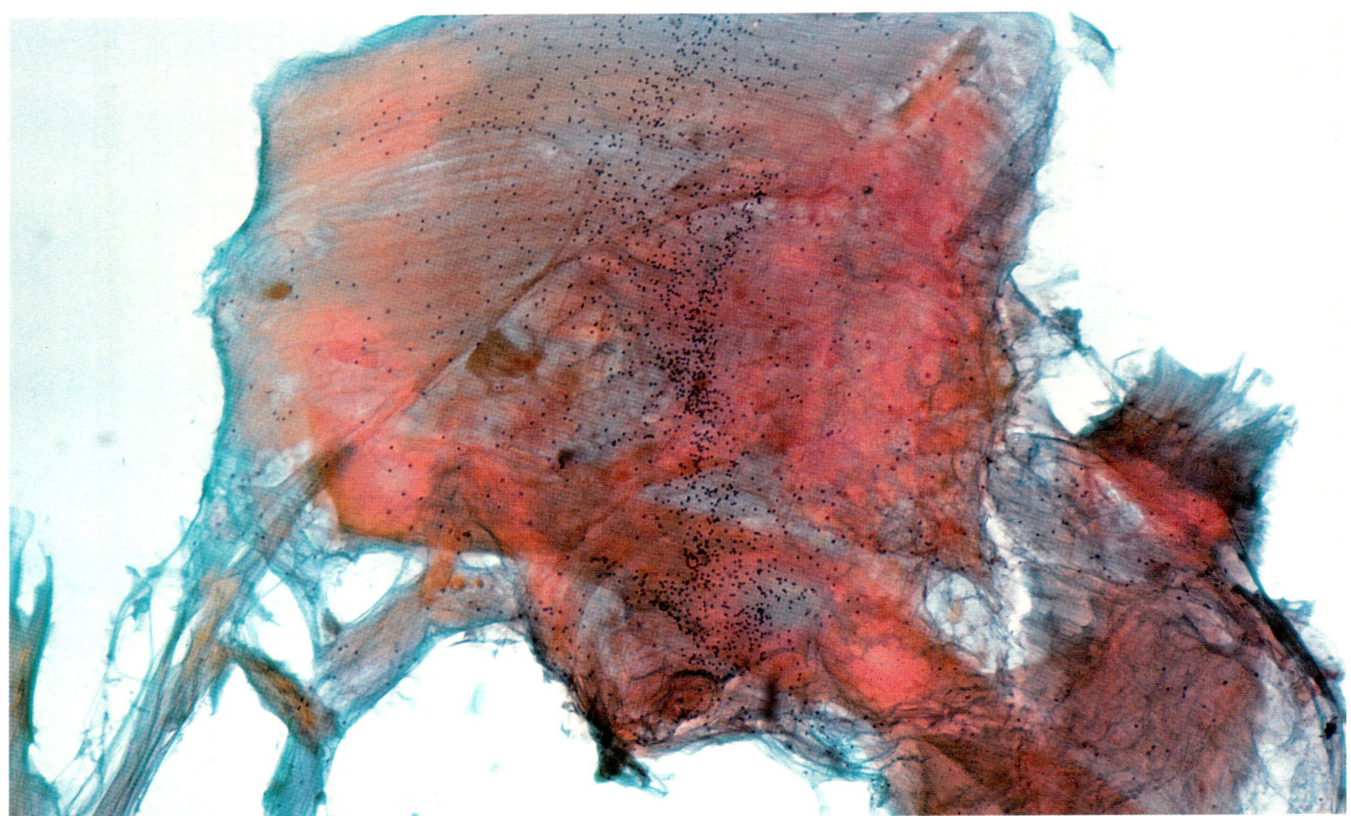

FIGURE 4–3 Numerous retained nuclei are noted in histologic section from the lens shown in Figure 4–2. (hematoxylin and eosin, × 125)

FIGURE 4–4 Intraocular washing from another case of congenital rubella shows retained nuclei. (Papanicolaou, × 125)

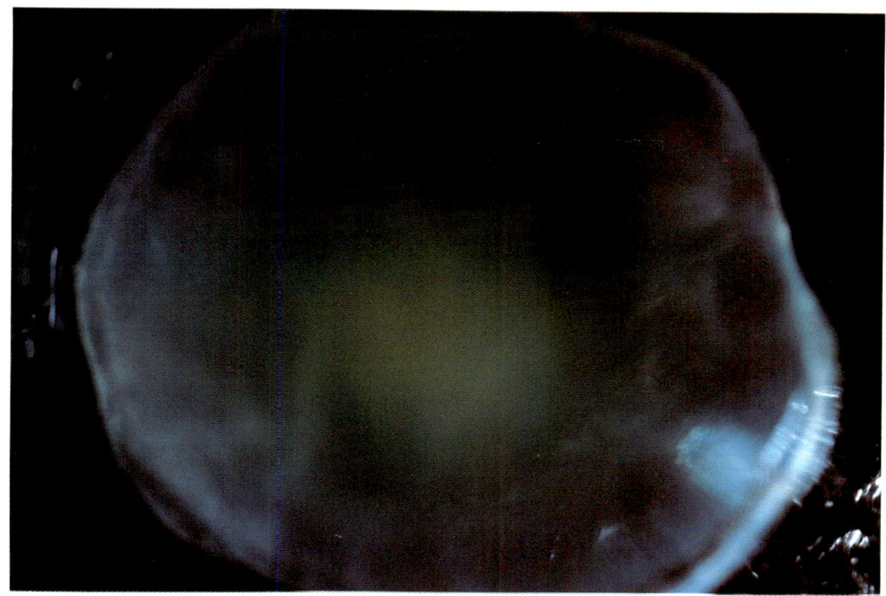

FIGURE 4–5 Gross photograph shows central yellowing indicative of nuclear sclerosis in an adult lens.

FIGURE 4–6 Photomicrograph shows melding of central lens fibers indicative of nuclear cataract. (hematoxylin and eosin, × 20)

FIGURE 4–7 Photomicrograph from cytospin preparation shows numerous eosinophilic bodies (Morgagnian globules) seen in cortical cataract. (hematoxylin and eosin, × 125)

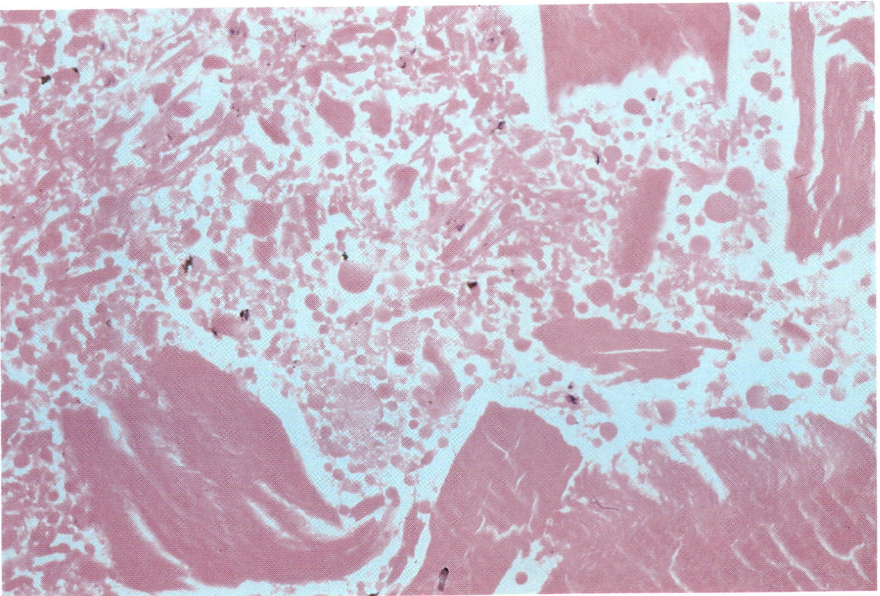

FIGURE 4–8 Gross photograph shows the round central opacity of a posterior subcapsular cataract (arrow). Pathology of aging-related cataracts. Reproduced with permission from Straatsma BR, Horwitz J, Gardner KM, Pettit TA et al. Ann Int Med 1985;102:82–92.

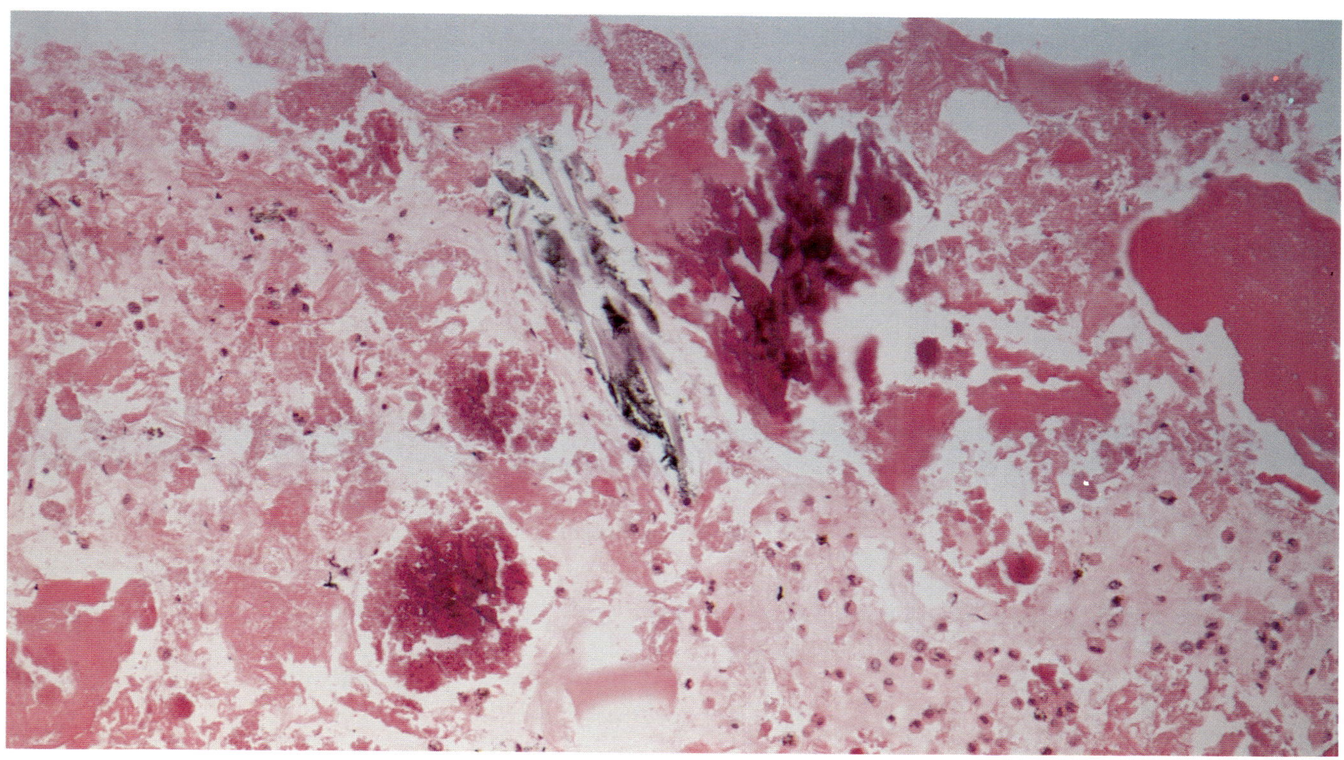

FIGURE 4–9 Cytospin preparations show the purple crystals indicative of calcification in a cataract. (hematoxylin and eosin, × 125)

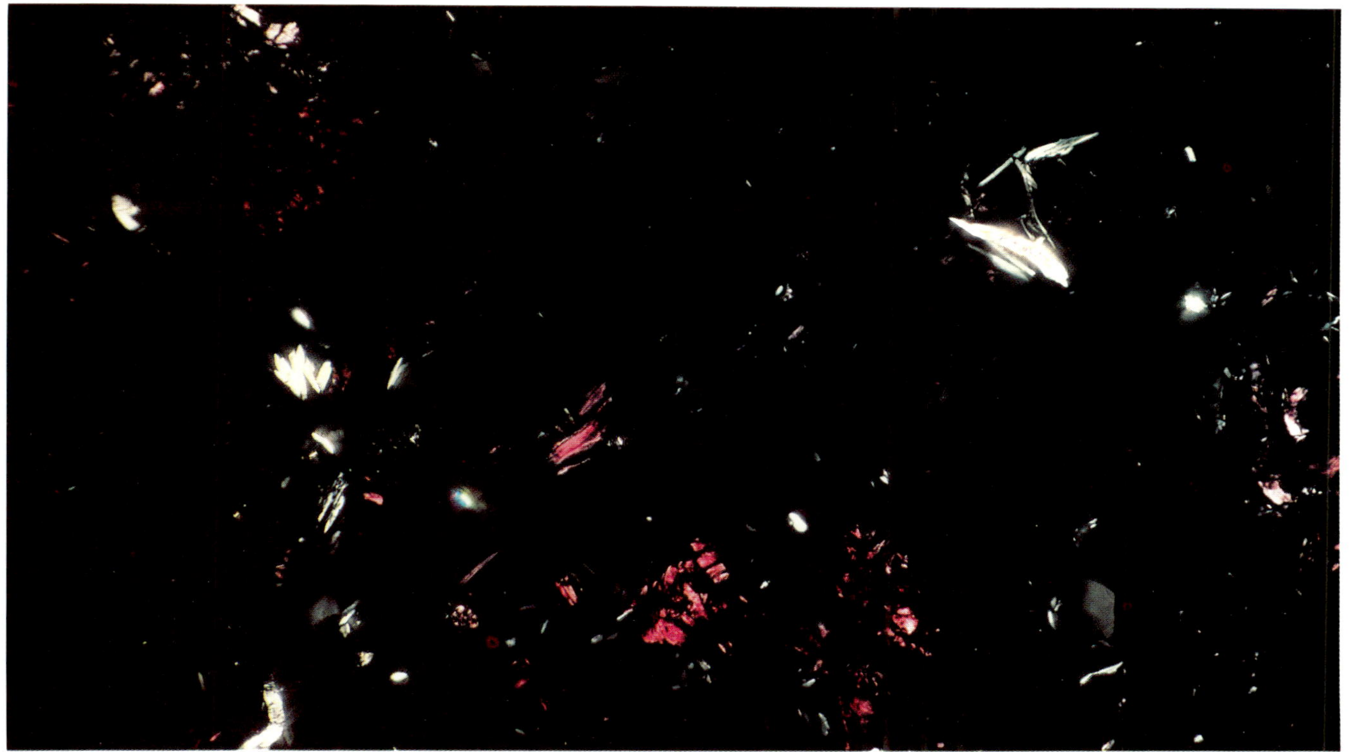

FIGURE 4–10 Polarized light displays the birefringence characteristic of calcification in a cataract. (May-Grünwald Giemsa, × 200)

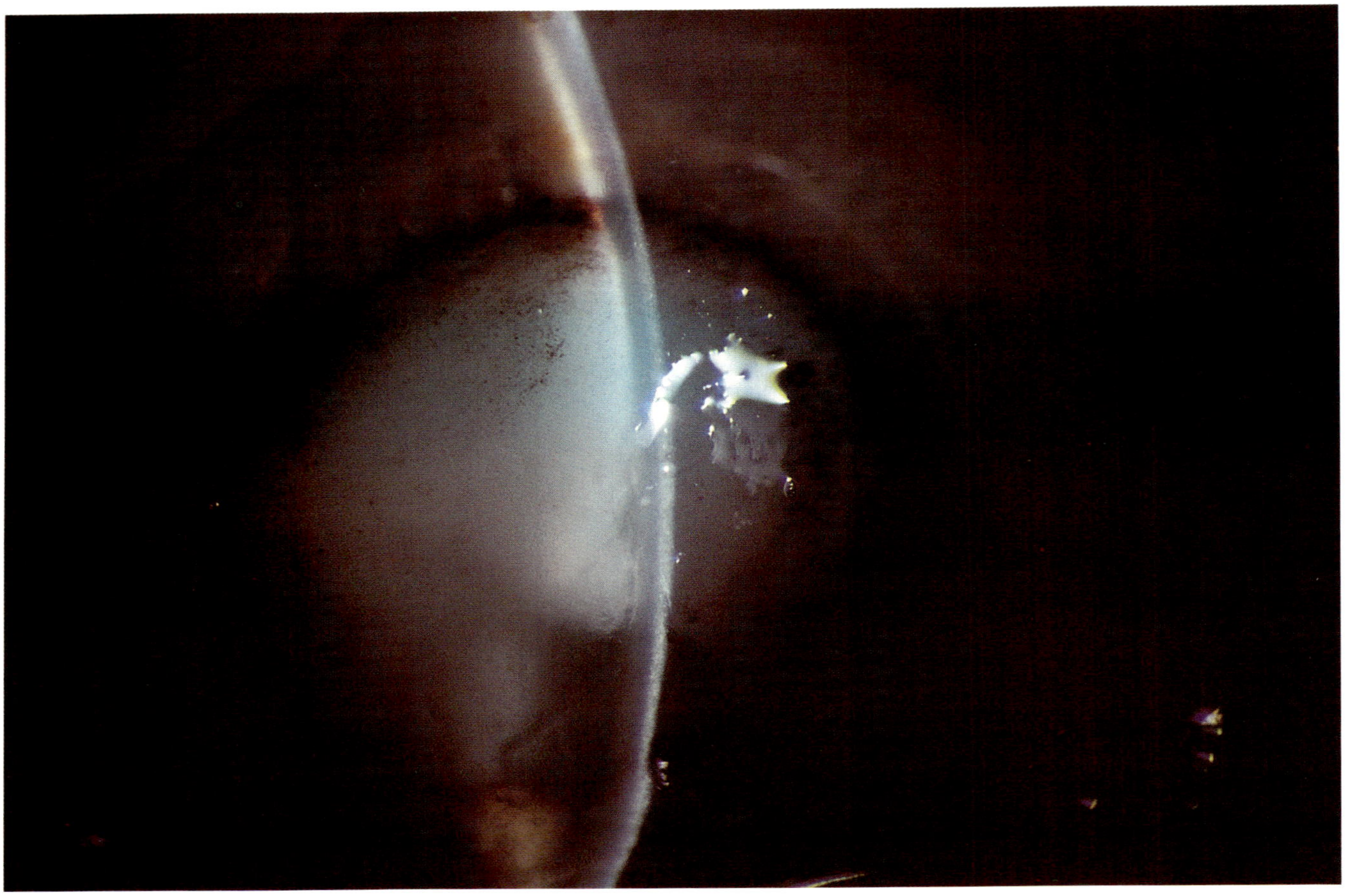

FIGURE 4–11 Clinical photograph taken three weeks after lens rupture associated with blunt ocular trauma. Intraocular pressure was markedly elevated. The lens capsule ruptured and lens material spilled into the anterior chamber and produced severe inflammation (phacogenic glaucoma). Photograph courtesy of Dr. James D. Shuler, Jules Stein Eye Institute.

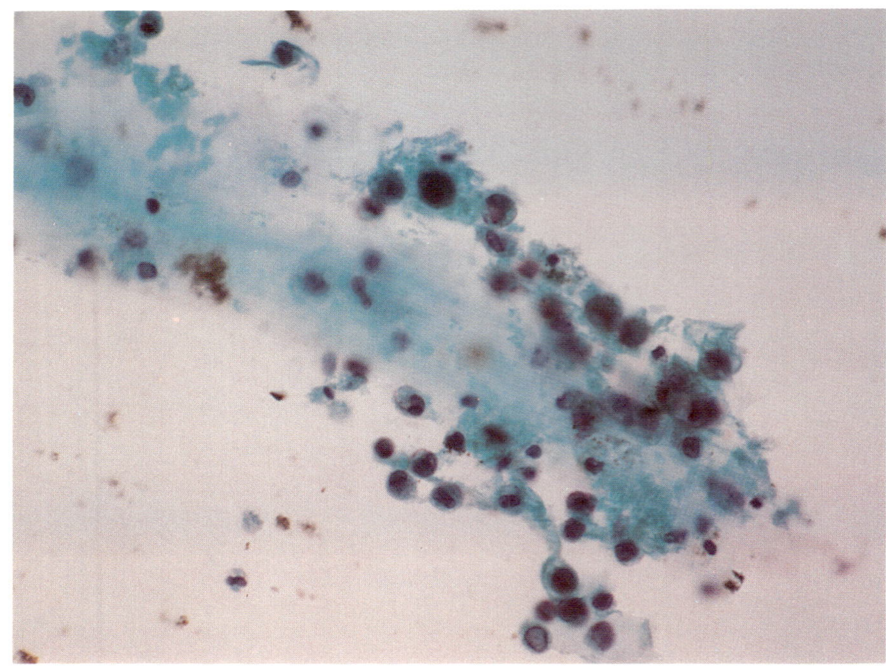

FIGURE 4–12 Numerous lens macrophages are evident from emulsification of lens after traumatic rupture. The patient had elevated intraocular pressure. (Papanicolaou, × 250)

PHACOANAPHYLACTOID AND PHACOLYTIC REACTIONS

Acute traumatic lens injury may result in a release of lens protein followed by a suppurative and granulomatous inflammatory response (Figure 4–11). Cytologically, the macrophages show ingested lens material (Figure 4–12).[12] Macrophages may obstruct the trabecular meshwork and produce elevated intraocular pressure (phacoanaphylactic glaucoma).[13] In hypermature cataracts, lens protein may leak through the capsule without evidence of trauma and produce a macrophage inflammatory response that may obstruct the trabecular meshwork (phacolytic glaucoma).[14] In addition, soluble lens proteins can obstruct aqueous outflow pathways and may be a factor in both of these lens-related glaucomas.[15]

REFERENCES

1. Boniuk M. Rubella and other intraocular viral diseases of infancy. Int Ophthalmol Clin 1972;12:1–137.

2. Yanoff M, Schaffer DB, Scheie HG. Rubella ocular syndrome—clinical significance of viral and pathologic studies. Trans Am Acad Ophthalmol Otolaryngol 1968;72:896–902.

3. Boniuk M, Zimmerman LE. Ocular pathology in the rubella syndrome. Arch Ophthalmol 1967;77:455–473.

4. Wolter JR, Insel PA, Willey EN, Brittain HP. Eye pathology following maternal rubella: a study of four children. J Pediat Ophthalmol 1966;3:29–35.

5. Zimmerman LE. The histopathologic basis for the ocular manifestations of the congenital rubella syndrome. Proc Inst Med Chicago 1967;26:170–192.

6. Zimmerman LE, Font RL. Congenital malformations of the eye: some recent advances in knowledge of the pathogenesis and histopathological characteristics. JAMA 1966; 196:684–692.

7. Hara J, Fujimoto F, Ishibashi T, Seguchi T, Nashimura K, et al. Ocular manifestations of the 1976 rubella epidemic in Japan. Am J Ophthalmol 1979;87:642–645.

8. Boger WP III. Late ocular complications in congenital rubella syndrome. Ophthalmology 1980;87:1244–1252.

9. Yanoff M, Fine BS. Ocular pathology, a text and atlas. Philadelphia: J.B. Lippincott, 1989.

10. Harding CV, Chylack LT Jr., Susan SR, Lo W-K, Bobrowski WF, et al. Calcium containing opacities in the human lens. Invest Ophthalmol Vis Sci 1983;24:1194–1202.

11. Zimmerman LE, Johnson FB. Calcium oxalate crystals within ocular tissue. Arch Ophthalmol 1958;60:372–382.

12. Goldberg MF. Cytological diagnosis of phacolytic glaucoma utilizing millipore filtration of the aqueous. Br J Ophthalmol 1967;51:847.

13. Yanoff M, Scheie HG. Cytology of human lens aspirate. Arch Ophthalmol 1968;80:166–170.

14. Flocks M, Littwin CS, Zimmerman LE. Phacolytic glaucoma. Arch Ophthalmol 1955;54:37–45.

15. Epstein DL, Jedziniak JA, Grant MW. Obstruction of aqueous outflow by lens particles and by heavy molecular weight soluble lens proteins. Invest Ophthalmol Vis Sci 1978;17:272–277.

CHAPTER

5

Abnormalities of the Iris, Ciliary Body, and Angle Structures

In this chapter, lesions of the iris, ciliary body, and anterior chamber angle that are important in ocular cytology are illustrated. There are only a few lesions of the ciliary body, iris, and angle that are frequently encountered in cytologic specimens. Infectious processes of the anterior chamber, iris, and ciliary body are presented in Chapter 8. Leukemias may involve the iris and shed cells into the aqueous, but are presented with other malignant neoplasms in Chapter 9. Choroidal melanomas are discussed in Chapter 9, but ciliary body melanomas are illustrated here.

HEMOGLOBIN SPHERULOSIS AND GHOST ERYTHROCYTES

Red blood cells that have hemolyzed and hemoglobin aggregates can be identified on cytologic examination from vitrectomy and aqueous specimens.[1] Ghost erythrocytes are red blood cells in which the bulk of the hemoglobin has been extruded, leaving behind small fragments of hemoglobin festooned along the cell membrane remnants. Ghost erythrocytes are important to identify because they may enter the anterior chamber angle, obstruct the trabecular meshwork, and produce ghost-cell glaucoma.[2] Ghost-cell glaucoma has been reported after vitreous hemorrhage associated with cataract extraction or diabetic retinopathy.[3]

Ghost cells appear on cytologic preparations as small ovoid to circular rings with central clear areas corresponding to the absence of hemoglobin. The hemoglobin is clumped at the periphery of the cell-forming Heinz bodies (Figure 5–1). Hemoglobin spherules are spheres of hemoglobin varying markedly in size, some of which are smaller than normal erythrocytes and others three to five times the diameter of a red blood cell (Figure 5–2). Like ghost cells, hemoglobin spherules may obstruct the trabecular meshwork and produce glaucoma. If ghost cells or hemoglobin aggregates are diagnosed by cytologic examination of an aqueous aspirate, vitrectomy may be indicated in aphakic patients to relieve glaucoma.[4]

PARS PLANITIS

Pars planitis is a condition in which cells and membranous veils are observed in the vitreous overlying the pars plana. This finding is termed "*snowbanking*."[5] The cause of the disease is unknown. It is usually bilateral and occurs in young individuals. The collection of vitreous cells may obscure vision. Occasionally, vitrectomy may be done if an infectious agent is sought or if the opacity is chronic and severely limits vision. Histologically, lymphocytes and macrophages are seen adjacent to hyper-plastic nonpigmented ciliary epithelium. In the chronic condition, membranes may form over the pars plana composed of collagen and scarce spindle cells (Figure 5–3). Perivascular sheathing of lymphocytes may also be seen (Figure 5–4). Cytologic preparations show lymphocytes and macrophages enmeshed in fibroglial membranous fragments (Figure 5–5). These findings are not specific, and the diagnosis requires careful clinical correlation.

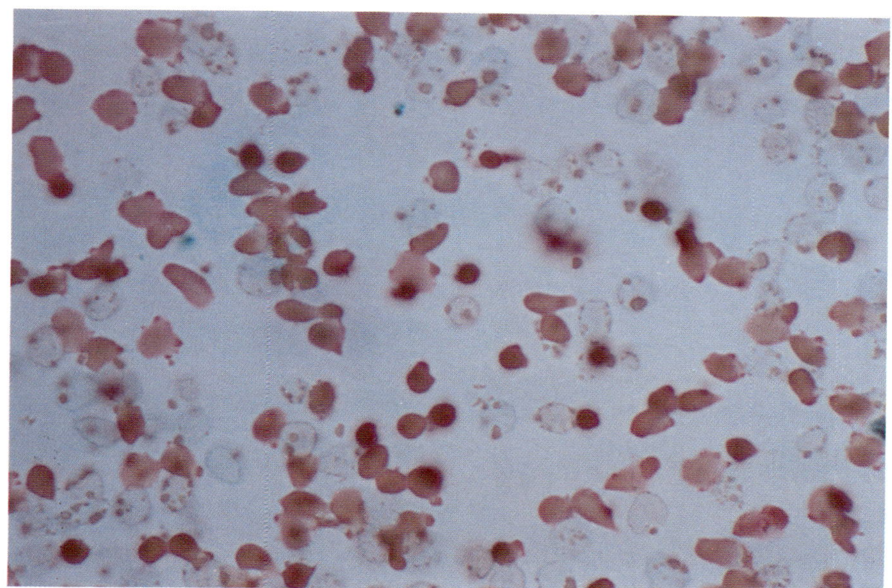

FIGURE 5–1 Photomicrograph demonstrates ghost erythrocytes with hemoglobin clumping at the cell periphery (Heinz body). Vitreous hemorrhage was the clinical diagnosis. (hematoxylin and eosin, × 480)

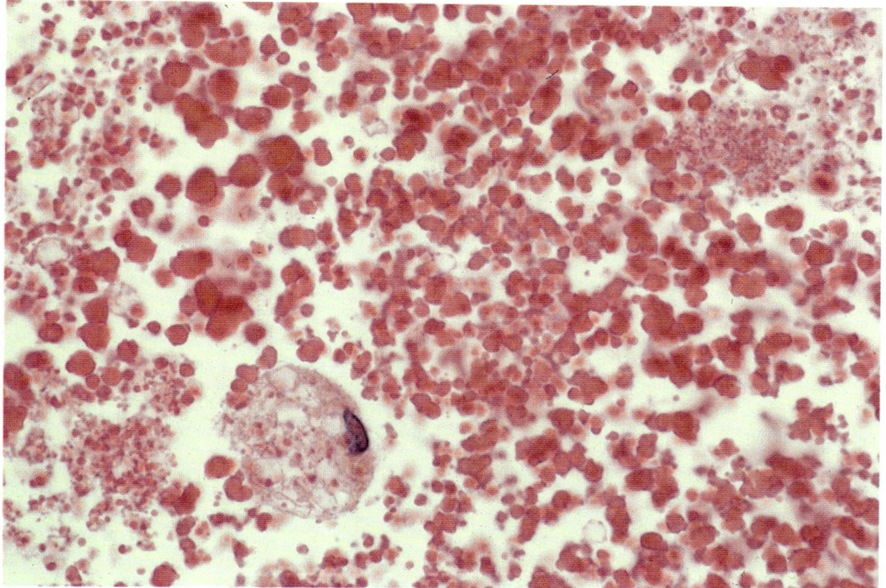

FIGURE 5–2 Photomicrograph shows variable-sized round fragments indicative of hemoglobin spherulosis. (hematoxylin and eosin, × 480)

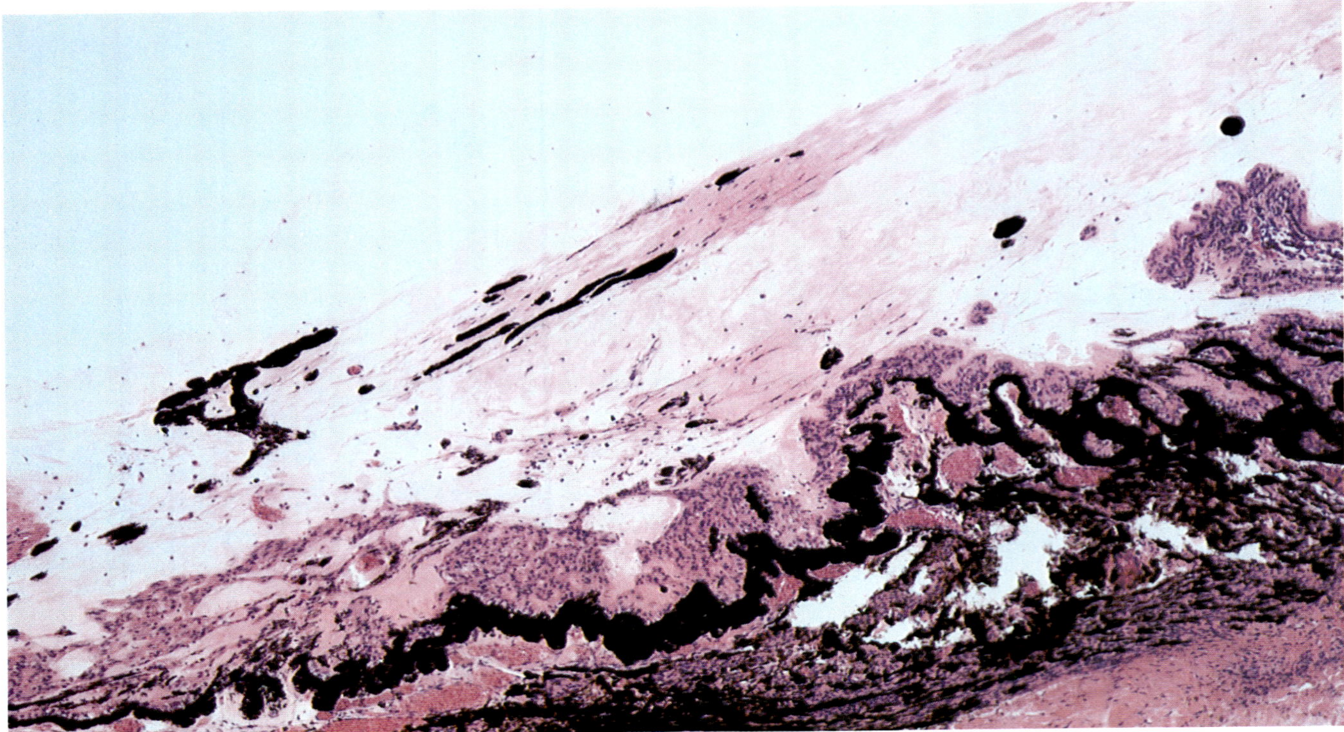

FIGURE 5–3 (ABOVE) Photomicrograph shows histologic sections from a patient with the clinical diagnosis of pars planitis. Lymphocytes are enmeshed in a fibroglial veil over the ciliary body. (hematoxylin and eosin, × 40)

FIGURE 5–4 (BELOW) Photomicrograph of the retina in pars planitis demonstrates lymphocytes cuffed around a nearly obliterated retinal blood vessel. (hematoxylin and eosin, × 300)

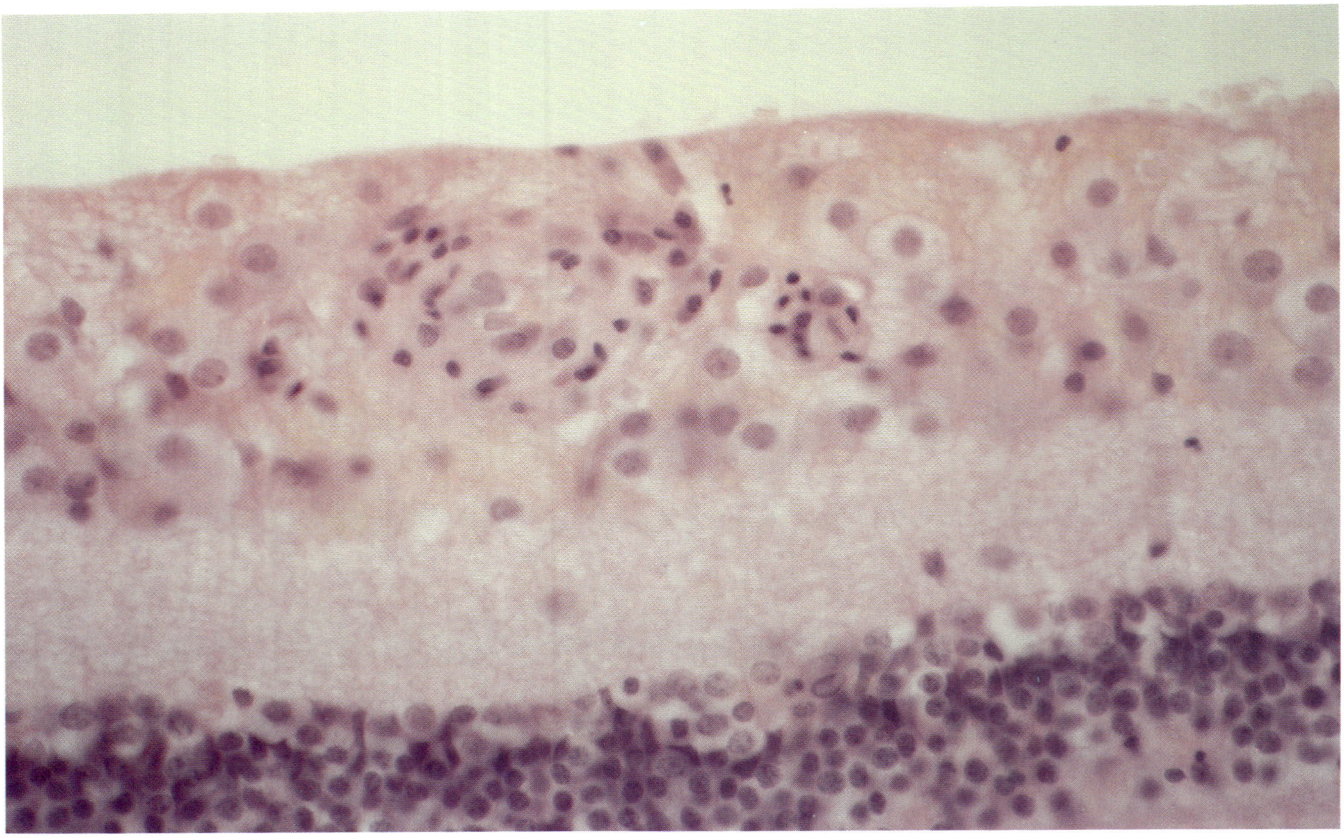

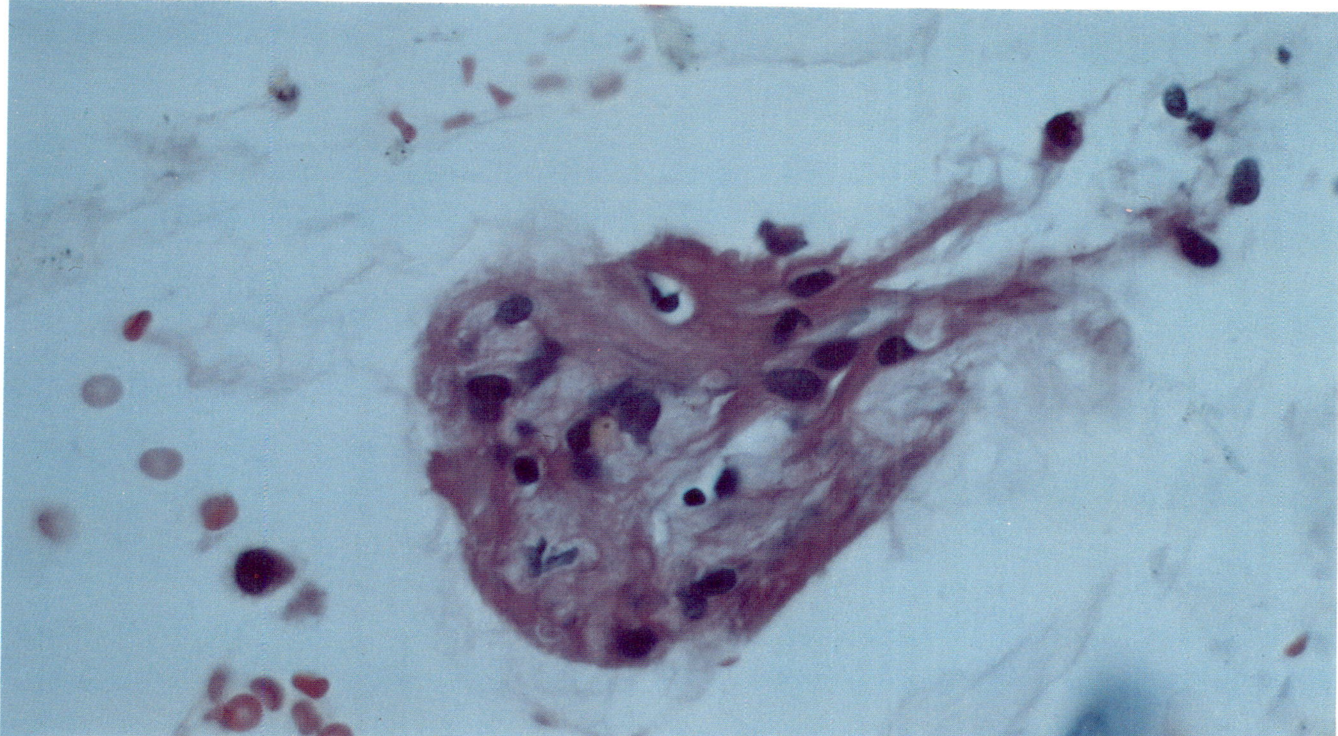

FIGURE 5–5 **Photomicrograph shows a fibroglial membranous tissue fragment surrounded by lymphocytes. (hematoxylin and eosin, × 250)**

CORONAL ADENOMA

The coronal adenoma, also called benign adenoma, Fuchs' adenoma, Fuchs' epithelioma, and epithelial hyperplasia, is the most common intraocular tumor.[6,7] These adenomas originate from the nonpigmented epithelium of the pars plicata (the coronal processes) (Figure 5–6). They occur more commonly in older individuals.[8] Histologic sections show nonpigmented ciliary epithelium embedded in a matrix of basement membrane material (Figure 5–7). Rarely, these tumors are clinically misdiagnosed as melanoma because they indent the iris and mimic a pigmented lesion.[9,10] Aspiration biopsy will distinguish melanoma from coronal adenoma. Fine needle aspiration of coronal adenomas has revealed cohesive groups of nonpigmented epithelium with bland nuclei and abundant cytoplasm clustered around extracellular matrix material (Figure 5–8).[11] Ultrastructural and immunohistochemical studies indicate that the extracellular matrix material contains Type IV collagen.[12]

LEIOMYOMA

Leiomyomas of the ciliary body are rare.[13-18] They presumably arise from ciliary muscle. The clinical diagnosis may be suggested by augmented transillumination. Electron microscopic identification of myofilaments and micropinocytotic vesicles differentiates these tumors from neurofibromas and schwannomas.

Mesectodermal leiomyomas of the ciliary body are extremely rare.[19] Fine needle aspiration characteristics of these tumors have not been reported previously. We reviewed a fine needle aspiration of one such tumor from the UCLA cytology archives. The aspirate shows tight clusters of spindle and oval cells in a fibrillary background. The nuclei are extremely bland with inconspicuous nucleoli (Figure 5–9). Sections reveal an amelanotic ciliary body mass with prominent vascularity, a fibrillary and myxoid background, and individual tumor cells that are both spindled and oval in shape (Figures 5–10 and 5–11). These findings suggest a neural tumor, yet electron microscopy shows smooth muscle differentiation.[20]

MELANOCYTOMA

Melanocytoma is a benign lesion found most commonly in blacks.[21,22] It may be found in any part of the uveal tract, but is generally present adjacent to the optic disc. Melanocytomas occasionally occur in the ciliary body.[23] The cells of a melanocytoma are large with abundant densely pigmented cytoplasm and round nuclei with small nucleoli (Figure 5–12).

FIGURE 5–6 Gross photograph shows a white, round coronal adenoma of the ciliary processes (corona ciliaris).

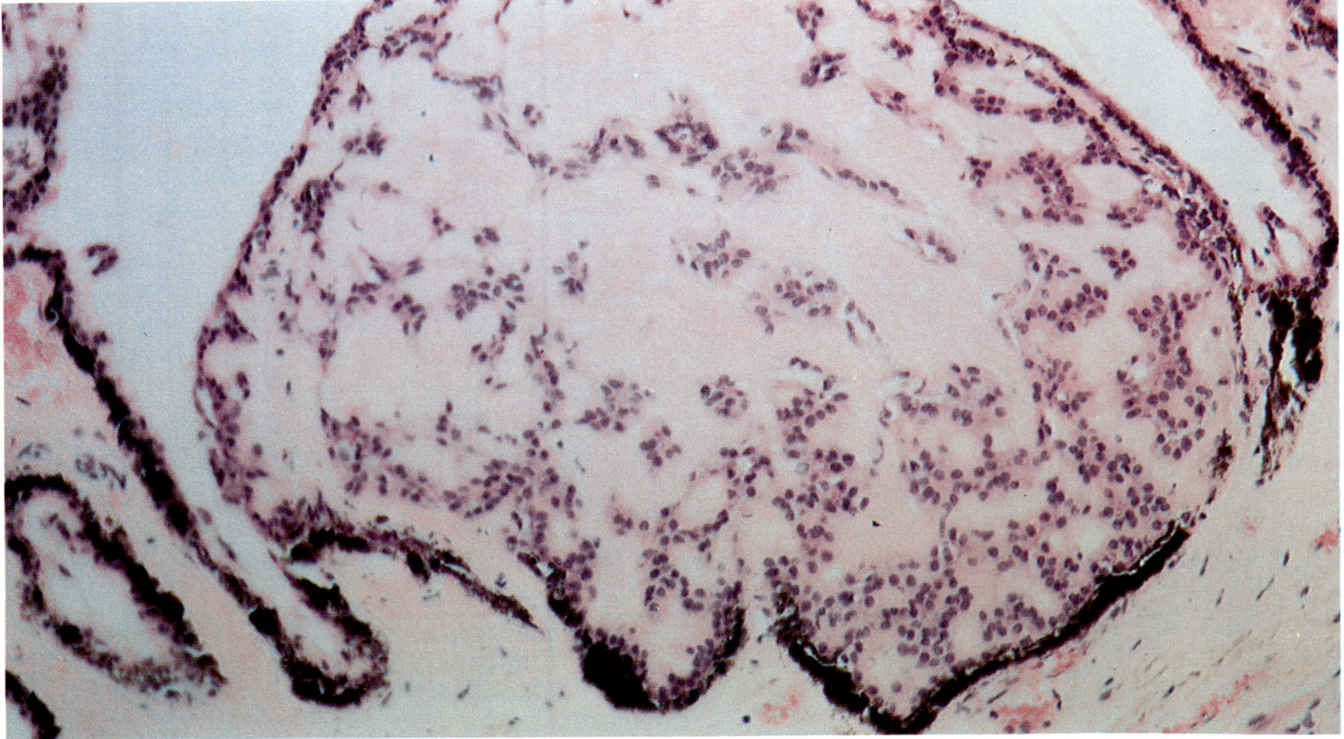

FIGURE 5–7 Histologic section shows a tumor (coronal adenoma) composed of nonpigmented ciliary epithelium embedded in extracellular matrix. (hematoxylin and eosin, × 125)

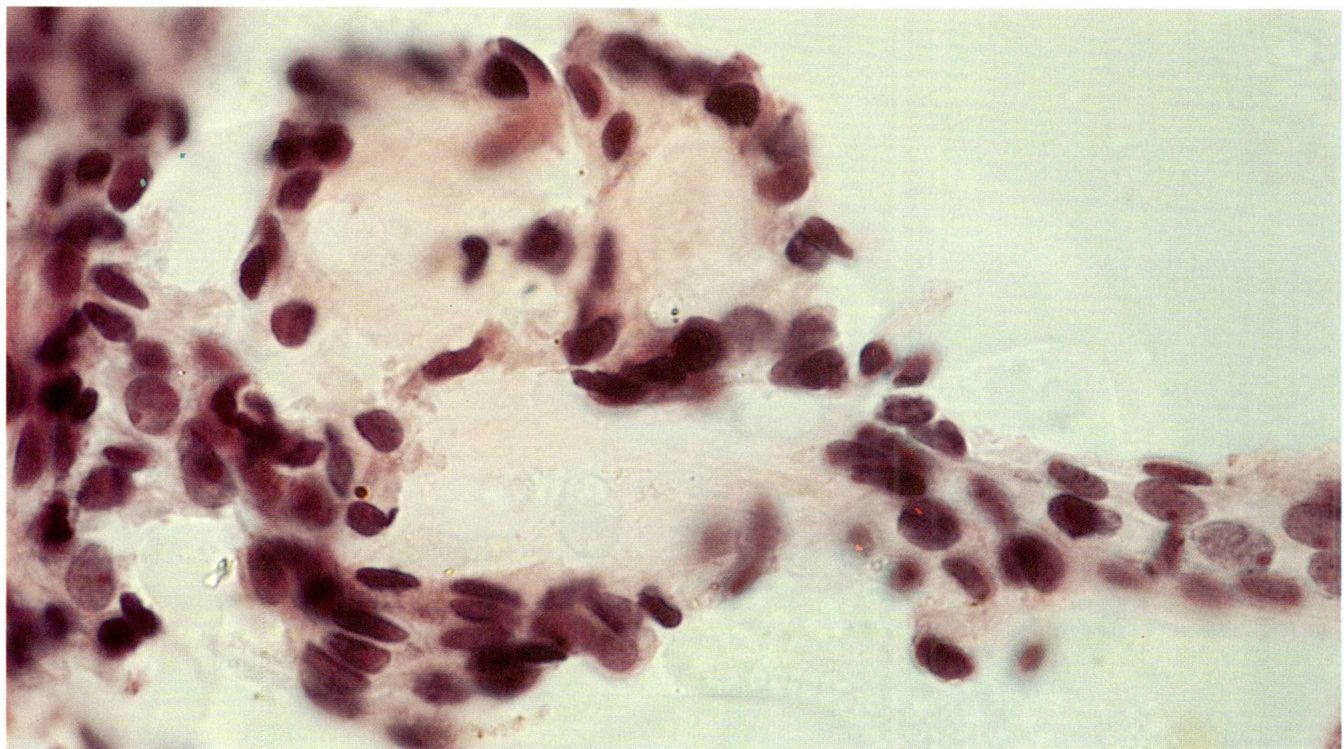

FIGURE 5–8 Fine needle aspirate of coronal adenoma shows nonpigmented epithelium surrounding extracellular basement membrane material. (hematoxylin and eosin, × 300)

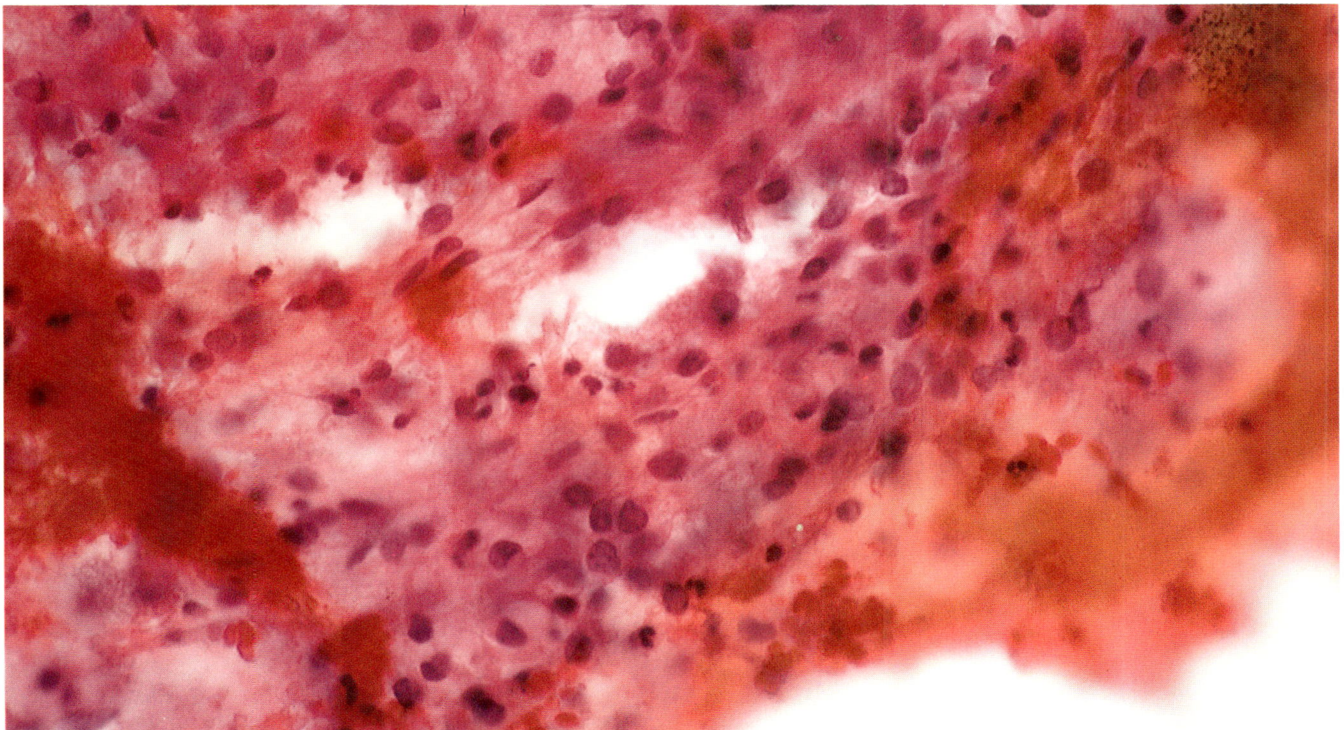

FIGURE 5–9 Photomicrograph of a fine needle aspirate from a 17-year-old female who complained of blurred vision for one year. A mass lesion with an overlying exudative retinal detachment was observed clinically. Ultrascund showed high internal reflectivity inconsistent with a diagnosis of melanoma. The lesion grew slightly over the next year. The fine needle aspirate shows a large amount of blood and a fibrillary background containing oval and spindle cells. (hematoxylin and eosin, × 480)

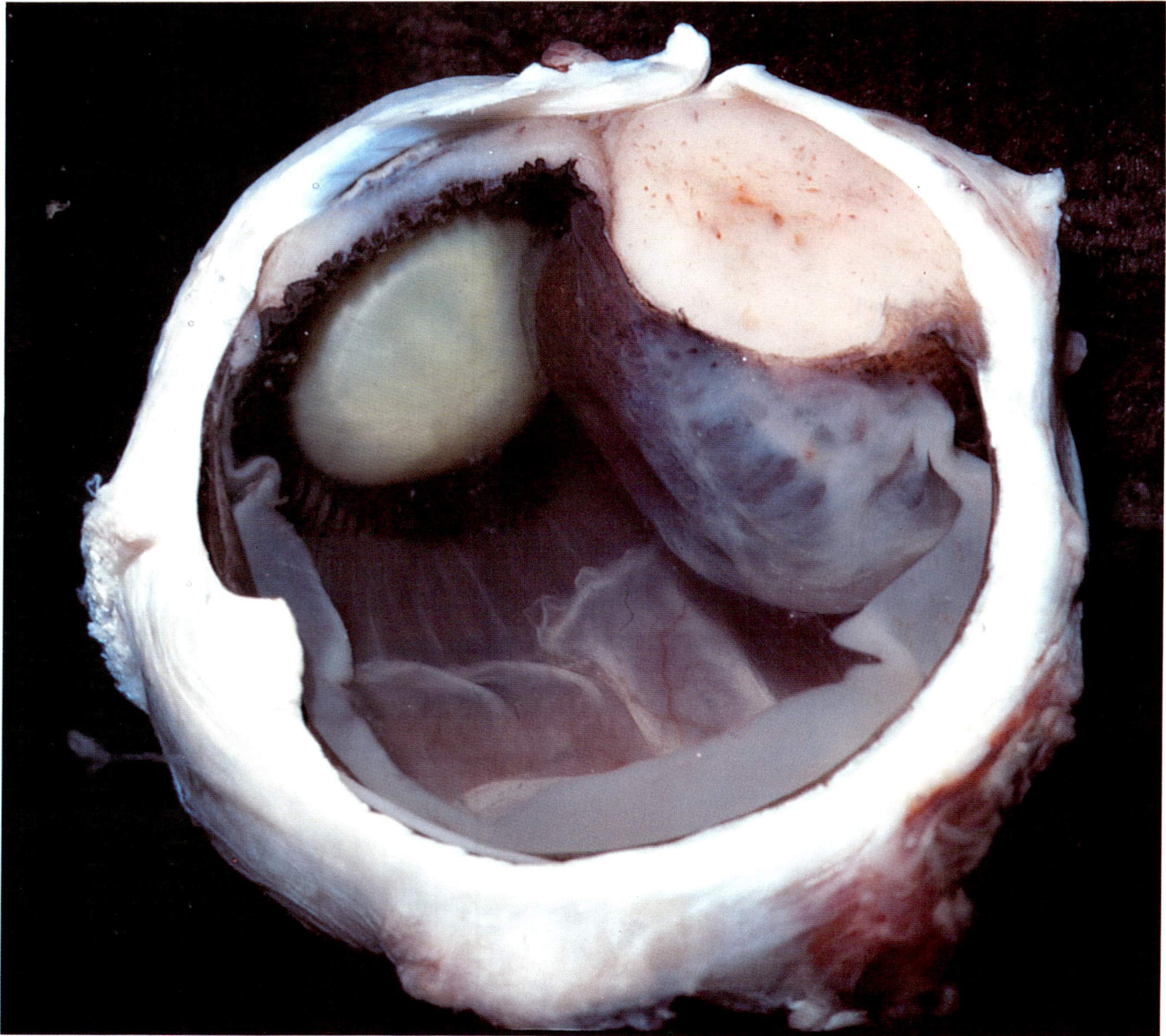

FIGURE 5–10 Gross photograph of the eye from which the smear in Figure 5–9 was taken shows a large ciliary body amelanotic mass impinging on the lens.

ADENOMA AND CARCINOMA OF THE PIGMENTED CILIARY EPITHELIUM

Adenomas of the pigmented ciliary epithelium may clinically mimic ciliary body melanomas because they are densely pigmented (Figure 5–13).[24–26] In general, they tend to be only locally invasive.[27] Immunohistochemical studies suggest that the tumor arises from the pigmented ciliary epithelium.[28] Histologically, the tumor replaces the ciliary body and is composed of large pigmented cells arranged in a nodular pattern (Figure 5–14). The cells surround cystic spaces which may appear to displace the nucleus (Figure 5–15). Cytologic findings include large cells with abundant intracellular and extracellular pigment, and round nuclei without prominent nucleoli.[29]

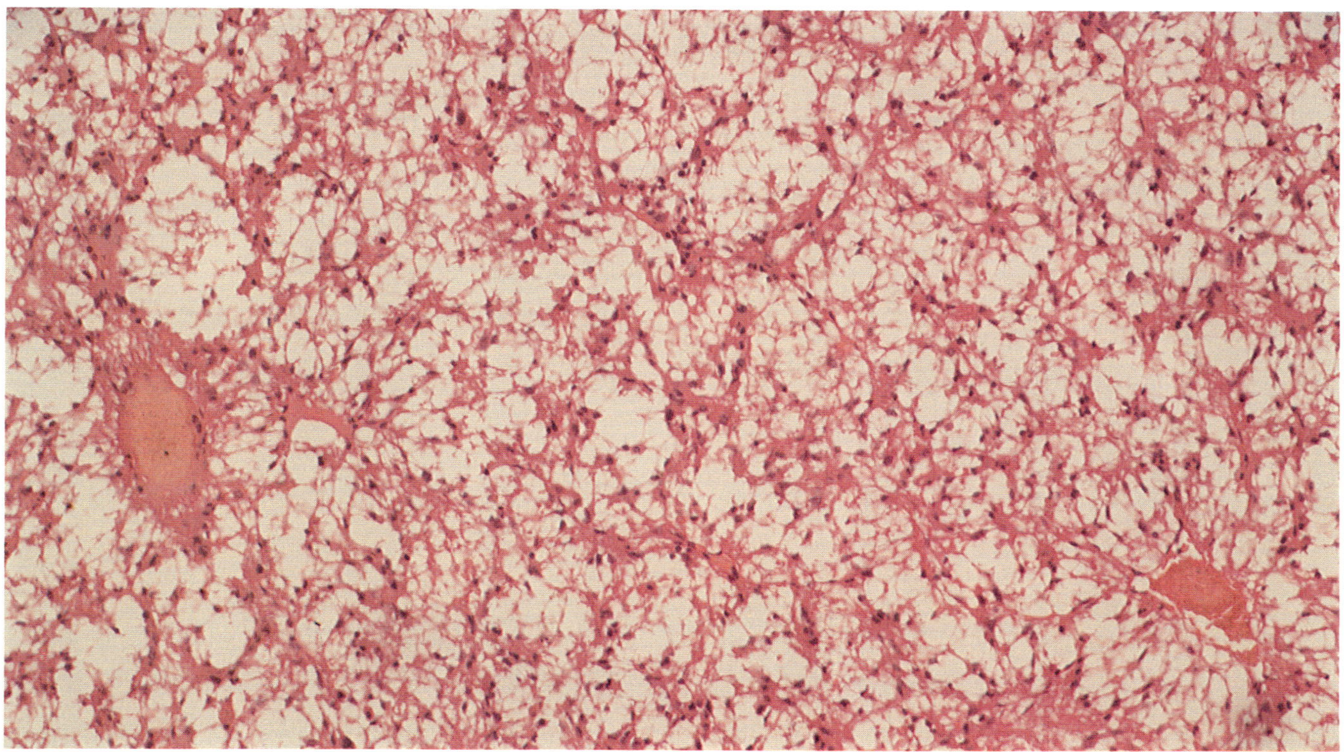

FIGURE 5–11 Photomicrograph reveals prominent blood vessels, a fibrillary background, myxoid areas, and both round and oval cells consistent with a mesectodermal leiomyoma. (hematoxylin and eosin, × 250)

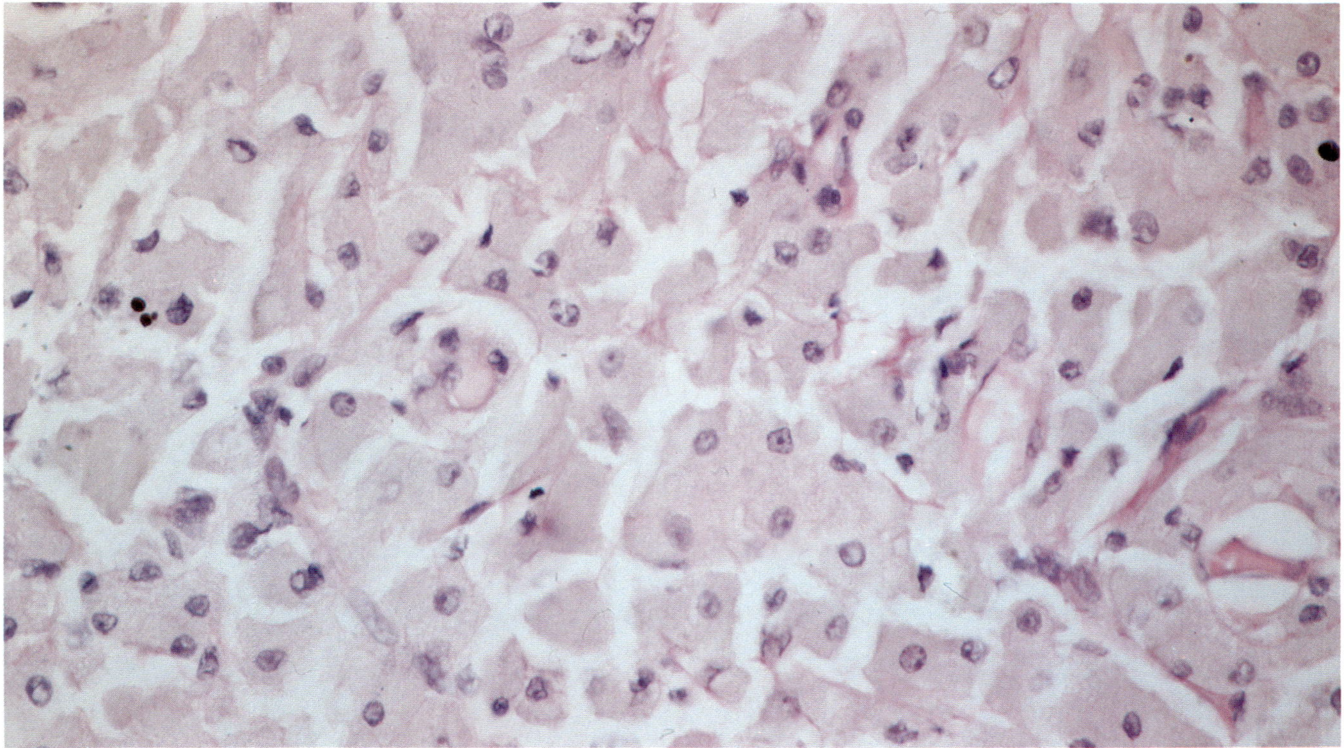

FIGURE 5–12 Histologic section of a melanocytoma shows large cells with pigmented cytoplasm and small round nuclei. (hematoxylin and eosin, × 300)

FIGURE 5–13 Gross photograph shows a pigmented ciliary body tumor macroscopically indistinguishable from melanoma.

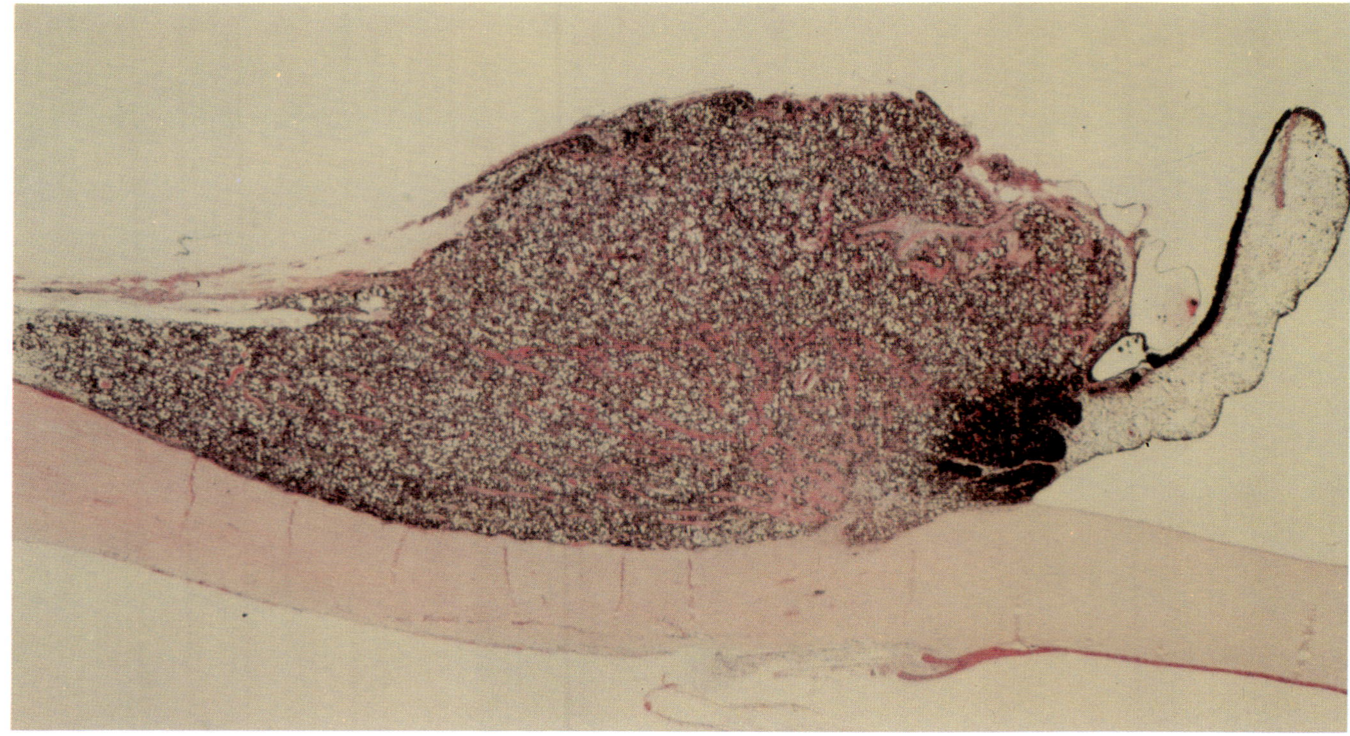

FIGURE 5–14 Histologic sections of tumor in Figure 5–13 show sheets of densely pigmented cells replacing the ciliary body and forming cystic spaces. (hematoxylin and eosin, × 40)

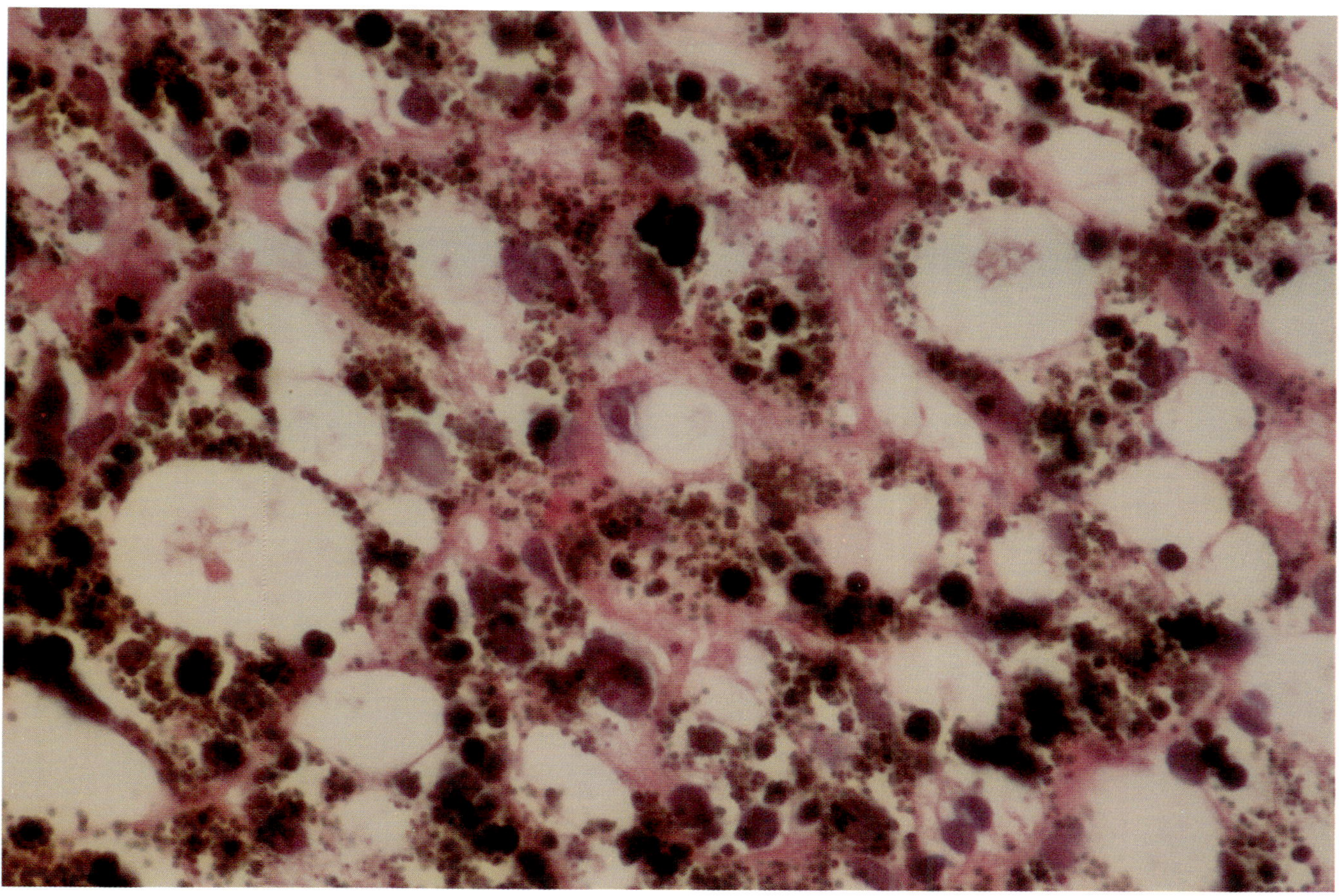

FIGURE 5–15 The cells have large pigmented granules, abundant cytoplasm, displaced nuclei, and single nucleoli. (hematoxylin and eosin, × 480)

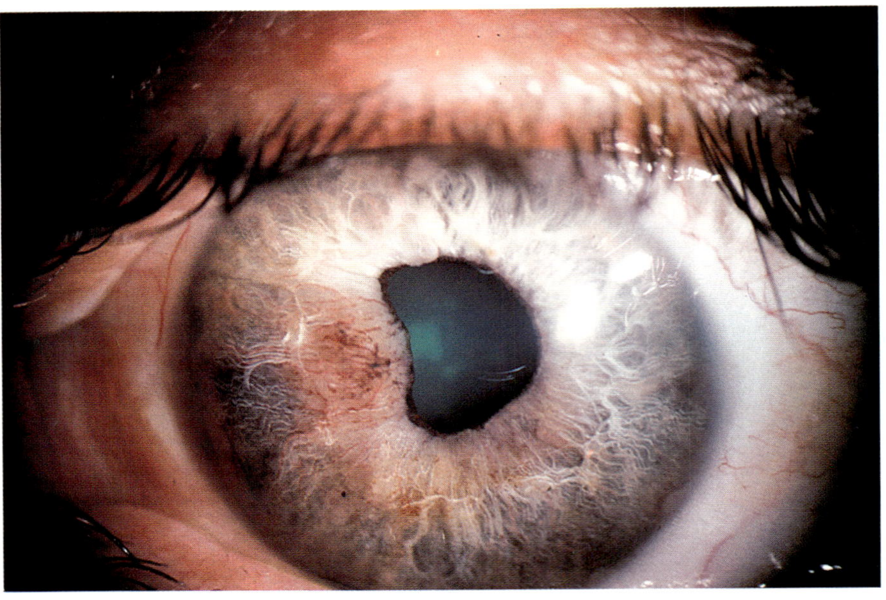

FIGURE 5–16 Clinical photograph from a 52-year-old male shows a vascular pigmented lesion located inferiorly in the iris. The lesion was diagnosed as an iris nevus for many years.

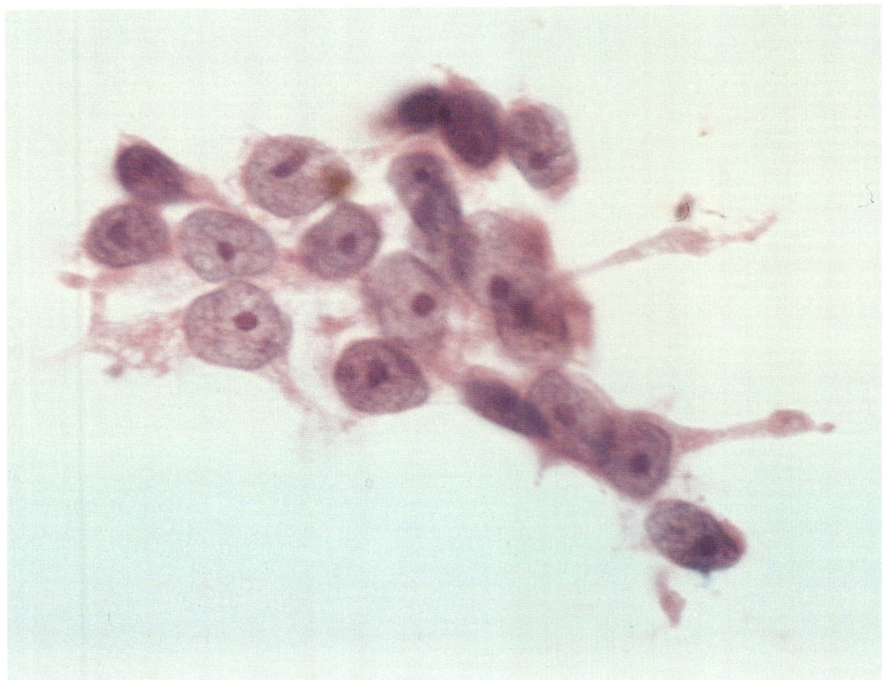

FIGURE 5–17 Fine needle aspiration from the patient shown in Figure 5–16 demonstrates clusters of cells with large round nuclei, prominent nucleoli, and pigmented cytoplasm (malignant melanoma). (hematoxylin and eosin, × 480)

IRIS AND CILIARY BODY MELANOMA

Melanomas in the eye most commonly occur in the choroid (see Chapter 9). However, about 3% to 16% (41% in children) of ocular melanomas arise in the iris and 9% to 10% arise in the ciliary body.[30–33] Clinically, iris melanomas are highly vascular, pigmented lesions that usually occur inferiorly in the iris.[34,35] The Callender classification of uveal melanomas as spindle or epithelioid does not apply to iris melanomas.[36–38] Iris melanomas have a much better prognosis than their ciliary body and choroidal counterparts.[39] Occasionally, malignant melanomas extend circumferentially about the ciliary body, yet clinically only a focal iris lesion is present (Figure 5–16). Determination of cell type in these cases may be helpful to document the presence of nucleoli suggesting a more atypical type of melanoma (Figure 5–17). Fine needle aspiration may be quite helpful in determining the presence of a ring melanoma of the ciliary body if the aspirate is performed away from the apparent lesion (Figures 5–18 and 5–19). Fine needle aspiration has been used to diagnose iris masses and, specifically, to help de-

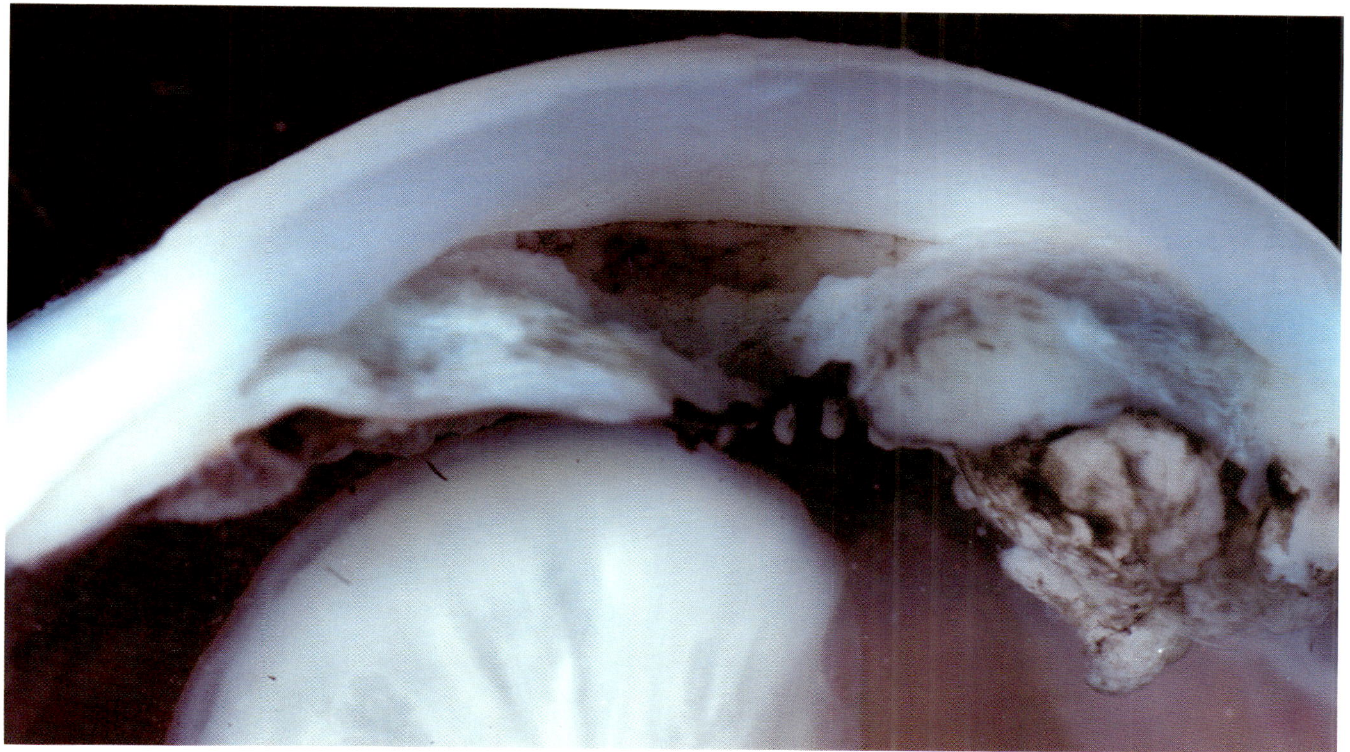

FIGURE 5–18 Gross photograph of the enucleated eye from the patient in Figures 5–6 and 5–7 shows a ring melanoma with extensive involvement of the ciliary body 230° from inferior mass.

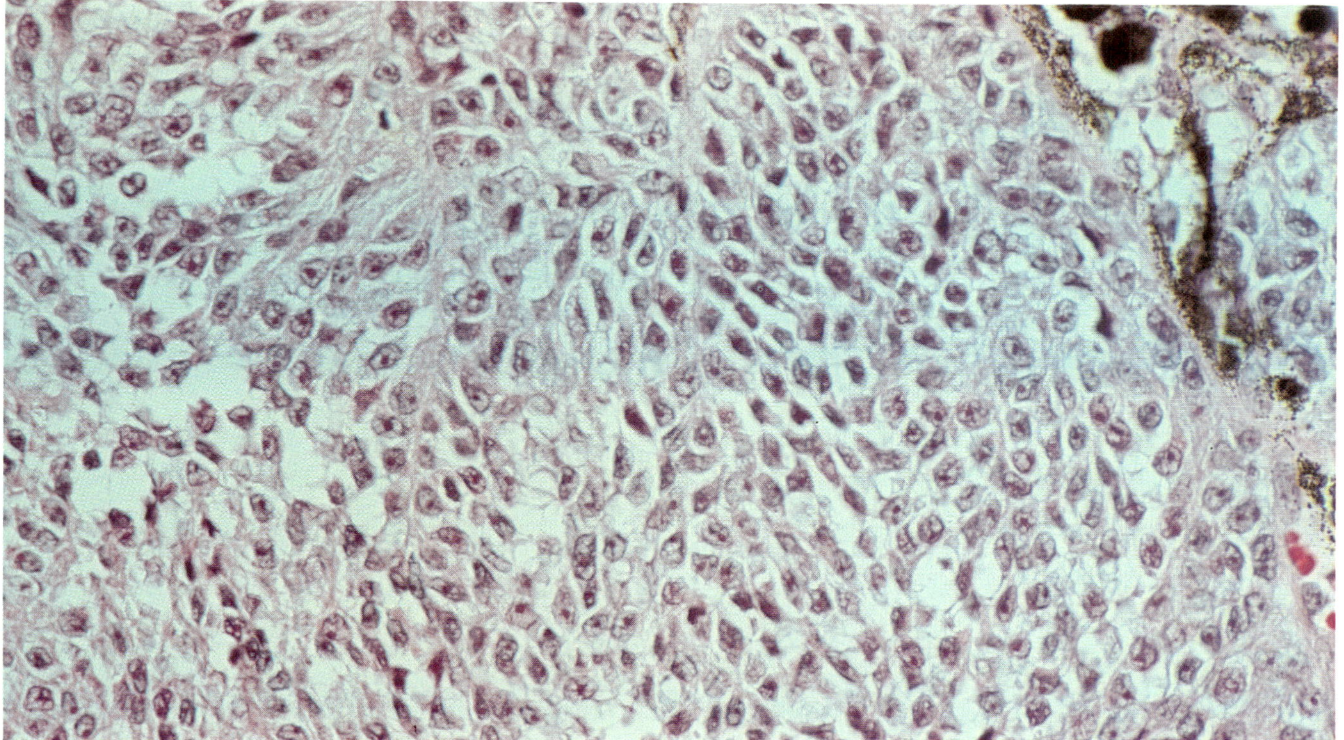

FIGURE 5–19 Histologic section confirms the presence of numerous tumor cells with prominent nucleoli and brown pigmented cytophom (hematoxylin and eosin, × 120)

termine if a pigmented lesion in the iris or ciliary body is focal, multiple, or diffuse.[40] However, on cytologic criteria alone, it is not possible to determine the original site of the melanoma.

REFERENCES

1. Grossniklaus HE, Frank KE, Farhi DC, Jacobs G, Green WR. Hemoglobin spherulosis in the vitreous cavity. Arch Ophthalmol 1988;106:961–962.

2. Campbell DG, Simmons RJ, Grant WM. Ghost cells as a cause of glaucoma. Am J Ophthalmol 1976;81:441–450.

3. Campbell DG, Essigmann EM. Hemolytic ghost cell glaucoma. Arch Ophthalmol 1979;97:2141–2146.

4. Brucker AJ, Michels RG, Green WR. Pars plana vitrectomy in the management of blood induced glaucoma with vitreous hemorrhage. Ann Ophthalmol 1978;10:1427–1437.

5. Pederson JE, Kenyon KR, Green WR, et al. Pathology of pars planitis. Am J Ophthalmol 1978;86:762–774.

6. Fuchs E. Anatomische miscellen. Albrecht Von Graefes Arch Klin Ophthalmol 1883;29:209.

7. Bateman JB, Foos RY. Coronal adenomas. Arch Ophthalmol 1979;97:2379–2384.

8. Iliff W, Green WL. The incidence and histology of Fuchs' adenoma. Arch Ophthalmol 1972;88:249–254.

9. Burch PG, Maumenee AE. Iridocyclectomy for benign tumors of the ciliary body. Am J Ophthalmol 1967;63:447–452.

10. Zaidman GW, Johnson BL, Salamon SM, Mondino BJ. Fuchs' adenoma affecting the peripheral iris. Arch Ophthalmol 1983;101:771–773.

11. Glasgow BJ. Intraocular fine needle aspiration biopsy of coronal adenomas. Diagn Cytopathol 1991;7:239–242.

12. Brown HH, Glasgow BJ, Foos RY. Coronal adenomas: ultrastructural and immunohistochemical features. Am J Ophthalmol 1991;112:34–40.

13. Blodi FC. Leiomyoma of the ciliary body. Am J Ophthalmol 1950;33:939–942.

14. Bonamour MM, Bonnet JC, Jambon M. Leiomyome du corps ciliaire; quelques considerations a propos du diagnostic et du traitement des tumeur benignes de l'iris et du corps ciliaire. Bull Soc Ophtalmol Fr 1957;7–8:482.

15. Calmettes L, Deodati F, Bec P. Leiomyome du corps ciliare. Bull Soc Ophtalmol Fr 1961;74:158–168.

16. Allen RA. Leiomyoma of ciliary body. Case presented at the Verhoeff Society, Washington, D.C., April 24–27, 1967.

17. Meyer SL, Fine BS, Font RL, Zimmerman LE. Leiomyoma of the ciliary body. Electron microscopic verification. Am J Ophthalmol 1968;66:1061–1068.

18. Calhoun FP Jr. Leiomyoma of the ciliary body. Case presented at the Verhoeff Society, Washington, D.C., April 26–29, 1976.

19. Jakobiec FA, Font RL, Tso MOM, Zimmerman LE. Mesectodermal leiomyoma of the ciliary body. A tumor of presumed neural crest origin. Cancer 1977;3 9:2102–2113.

20. Jakobiec FA, Iwamoto T. Mesectodermal leiomyoma of the ciliary body associated with a nevus. Arch Ophthalmol 1978;96:692–695.

21. Zimmerman LE. Melanocytes, melanocytic nevi, melanocytomas. Invest Ophthalmol 1965;4:11–40.

22. Reidy JJ, Apple DJ, Steinmetz RL, et al. Melanocytoma nomenclature, pathogenesis, natural history and treatment. Surv Ophthalmol 1985;29:319–327.

23. Frangieh GT, El Baba F, Traboulsi El, et al. Melanocytoma of the ciliary body: presentation of four cases and review of nineteen reports. Surv Ophthalmol 1985;29:328–334.

24. Streeten BF, McGraw JL. Tumor of the ciliary pigment epithelium. Am J Ophthalmol 1972;74:420–429.

25. Wilensky JT, Holland MG. A pigmented tumor of the ciliary body. Arch Ophthalmol 1974;92:219–220.

26. Chang M, Shields JA, Wachtel DL. Adenoma of the pigment epithelium of the ciliary body simulating a malignant melanoma. Am J Ophthalmol 1979;88:40–44.

27. Anderson SR. Medulloepithelioma of the retina. In: Zimmerman LE, ed. Tumors of the eye and adnexa. Int Ophthalmol Clin 1962;2:483–506.

28. Leib WE, Shields JA, Eagle RC, Kwa D, Shields CL. Cystic adenoma of the pigmented ciliary epithelium. Ophthalmology 1990;97:1489–1493.

29. Char DH, Miller TR, Crawford JB. Cytopathologic diagnosis of benign lesions simulating choroidal melanomas. Am J Ophthalmol 1991;112:70–75.

30. Jensen OA. Malignant melanoma of the human uvea. Acta Ophthalmol (Suppl.) 1963;75:17-215.

31. Raivio I. Uveal melanomas in Finland; an epidemiological, clinical, histological and prognostic study. Acta Ophthalmol (Suppl.) 1977;133:1–64.

32. Apt L. Uveal melanomas in children and adolescents. Int Ophthalmol Clin 1963;2:403–410.

33. Holland G. Clinical features and pathology of pigment tumours of the iris. Klin Monatsbl Augenheilkd 1967;150:359–370.

34. Rones B, Zimmerman LE. The prognosis of primary tumors of the iris treated by iridectomy. Arch Ophthalmol 1958;60:193–205.

35. Shields JA, Sandborn GE, Augsburger JJ. The differential diagnosis of malignant melanoma of the iris. A clinical study of 200 patients. Ophthalmology 1983;90:716–720.

36. Callender GR. Malignant melanotic tumors of the eye. A study of histologic types in 111 cases. Trans Am Acad Ophthalmol Otolaryngol 1931;36:131.

37. Jakobiec FA, Silbert G. Are most iris "melanomas" really nevi? Arch Ophthalmol 1981;99:2117–2132.

38. Arentsen JJ, Green WR. Melanoma of the iris: report of 72 cases treated surgically. Ophthalmic Surg 1975;6:23–37.

39. Ashton N. Primary tumours of the iris. BR J Ophthalmol 1964;48:650–668.

40. Char DH, Crawford JB, Gonzales J, Miller T. Iris melanoma with increased intraocular pressure. Differentiation of focal solitary tumors from diffuse or multiple tumors. Arch Ophthalmol 1989;107:548–551.

Chapter

6

Abnormalities of the Vitreous Body

Most abnormalities of the vitreous body (vitreous) are too subtle to be diagnosed by intraocular washings alone. These subtle changes are diagnosed by careful biomicroscopic examination in the living patient or with the dissecting microscope in the enucleated specimen. However, there are a few important conditions, such as amyloid, silicone retinopathy, and asteroid hyalosis, that are evident by examination of extracted vitreous. Other conditions, such as persistent hyperplastic primary vitreous, are considered in the clinical diagnosis of malignant tumors in children and it is important to be aware of their existence. Vitreous hemorrhage and hemoglobin spherulosis are presented in Chapter 5.

PERSISTENT HYPERPLASTIC PRIMARY VITREOUS

Persistent hyperplastic primary vitreous (PHPV) is a congenital condition most commonly unilateral in which there is persistent hyaloid vasculature and mesenchymal tissue from the embryonic primary vitreous in a microphthalmic eye.[1-3] Fibrovascular tissue and mesenchymal tissue are attached laterally to elongated and centrally dislocated ciliary processes (Figures 6–1 and 6–2). The lens-iris diaphragm may be anteriorly displaced and produce secondary angle-closure glaucoma. Cartilage, mature adipose tissue, and smooth muscle may arise from primitive mesenchymal tissue behind the lens.[4-6] There is frequently an accompanying total exudative retinal detachment (Figure 6–1). Clinically, the patient usually has a white pupillary reflex (leukocoria), poor vision, and a small eye. PHPV can be confused with retinoblastoma, although existence of leukocoria at birth and microphthalmia is highly suggestive of PHPV. If retinoblastoma is a serious consideration in the differential diagnosis or if surgical repair is considered for PHPV, preoperative fine needle aspiration or intraocular washing may be used. Smears show spindle cells and retinal tissue (Figure 6–3). Unlike retinoblastoma, the cells of PHPV are usually more cohesive and do not exhibit extensive necrosis.

ASTEROID HYALOSIS

Asteroid hyalosis is a disorder in which white-yellow smooth refractile spheres are suspended in the collagenous framework of the vitreous body. It is often unilateral, but can be bilateral. It is more common in older people.[7] The opacities are composed of complex lipids, calcium, and phosphorus.[8-10] Histochemically, the matrix of the opacities stain with periodic acid-Schiff, alcian blue, Sudan black B, and oil red O.[11] Clinically, asteroid hyalosis is often asymptomatic and associated with good visual acuity. However, it may be very dense and it may obscure vision.[12,13] In addition, the poor funduscopic view occasionally prevents adequate laser photocoagulation in patients with concomitant retinopathy (Figure 6–4). Vitrectomy may be indicated to improve vision or to improve the view of the fundus for the surgeon.[14] Cytologic specimens reveal amorphous round bodies that are birefringent in polarized light (Figures 6–5 and 6–6). There may be an associated foreign-body reaction.[10]

The cause of asteroid hyalosis is not known. Even though lipid is present in the yellow particles, there is no clear association with diabetes and hypercholesterolemia.[15-17]

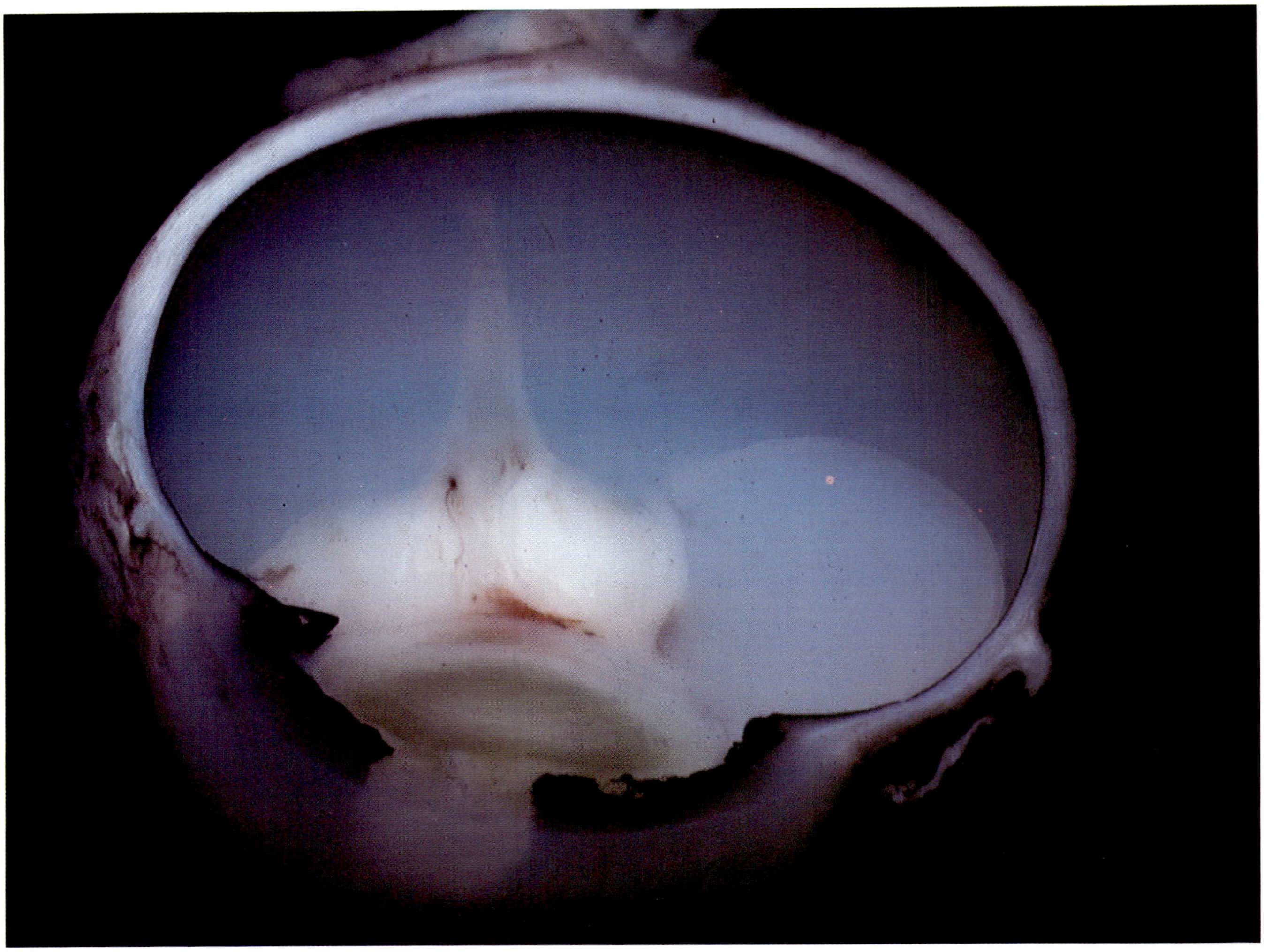

FIGURE 6–1 Gross photograph of a patient with PHPV demonstrates the small eye, complete exudative retinal detachment, and the mass of mesenchymal and retrolental retinal tissue.

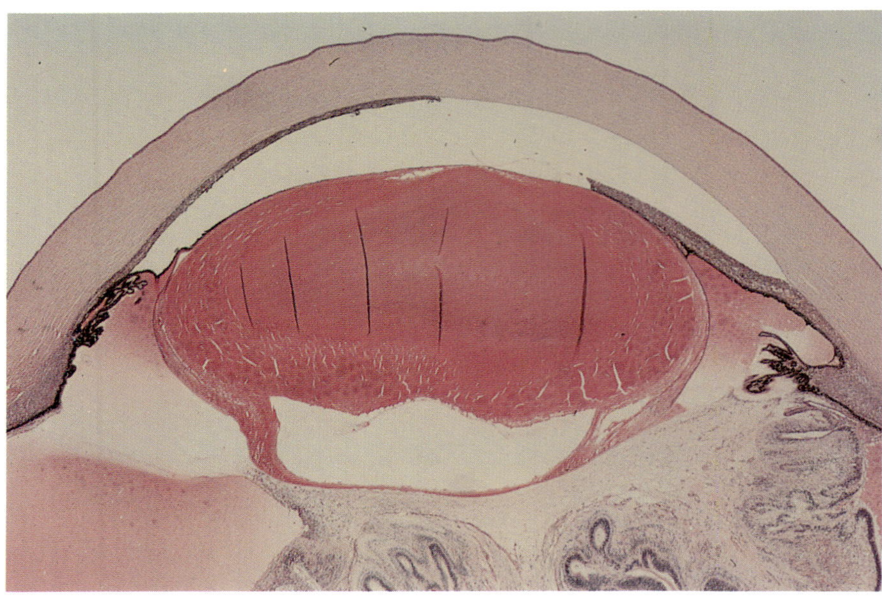

FIGURE 6–2 Histologic section of the eye shown in Figure 6–1 shows elongated ciliary processes, anterior displacement of the lens-iris diaphragm, and retrolental fibrovascular tissue and mesenchymal tissue. (hematoxylin and eosin, × 10)

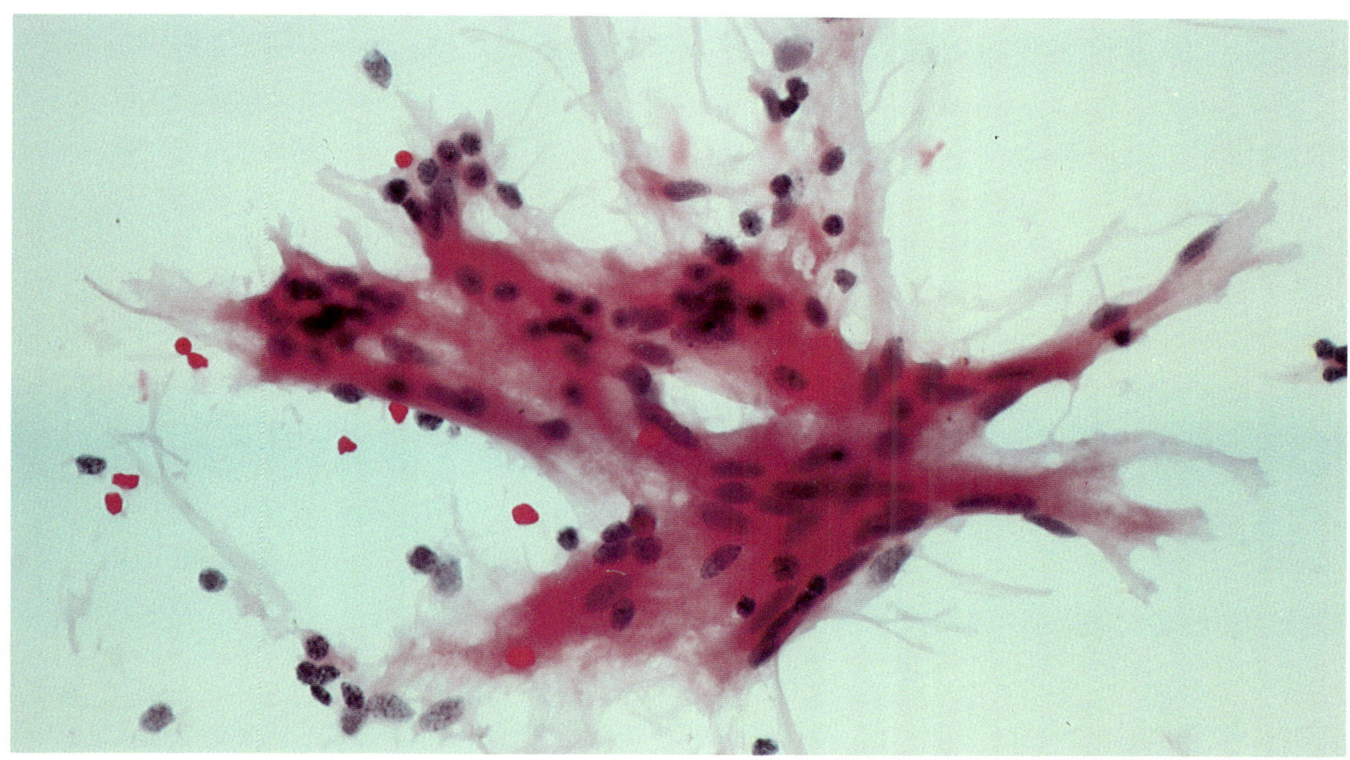

FIGURE 6–3 Photomicrograph shows spindle cells with elongated cytoplasmic processes and round cells. In some areas the round cells had an organoid pattern. (hermatoxylin and eosin, × 250)

FIGURE 6–4 Gross photograph shows numerous refractile yellow concretions in the vitreous body (asteroid hyalosis).

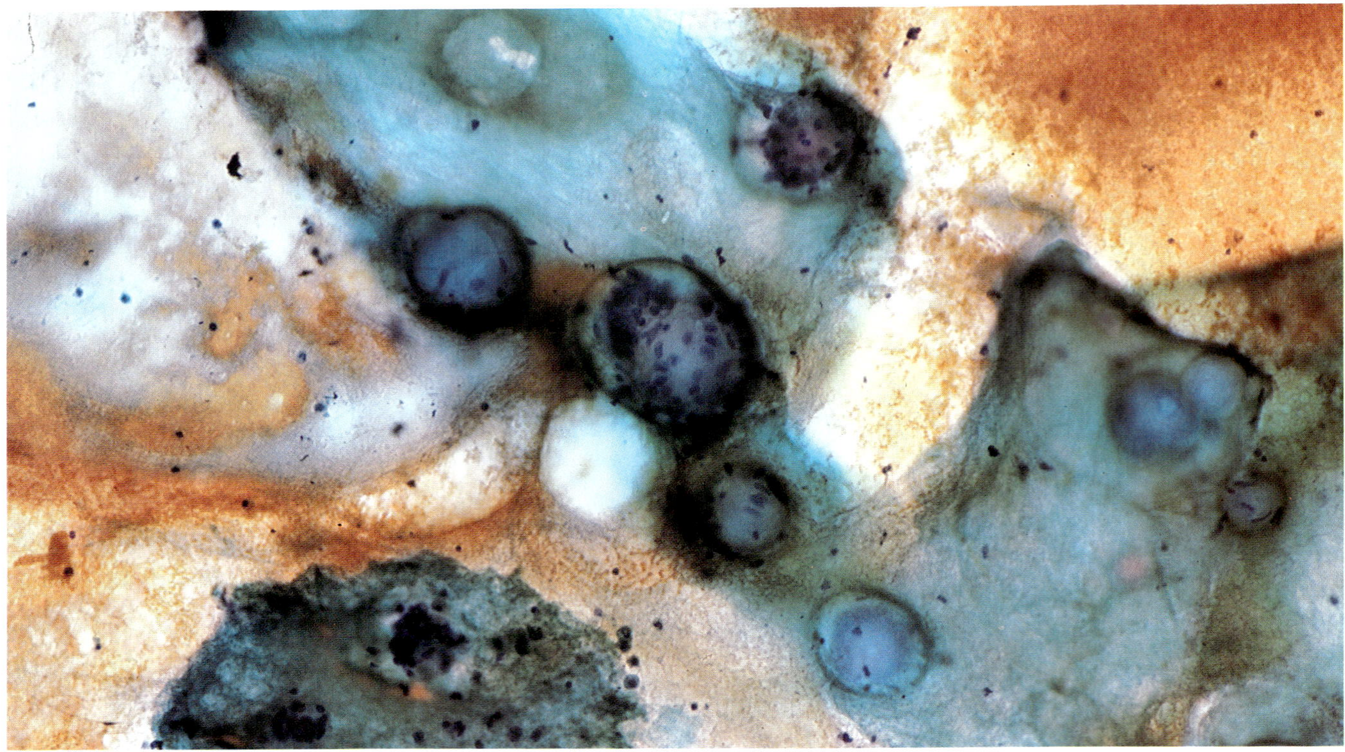

FIGURE 6–5 Photomicrograph shows numerous round amorphous concretions (asteroid bodies). (Papanicolaou, × 250)

FIGURE 6–6 Polarized light displays the birefringence seen with asteroid bodies. (Papanicolaou, × 250)

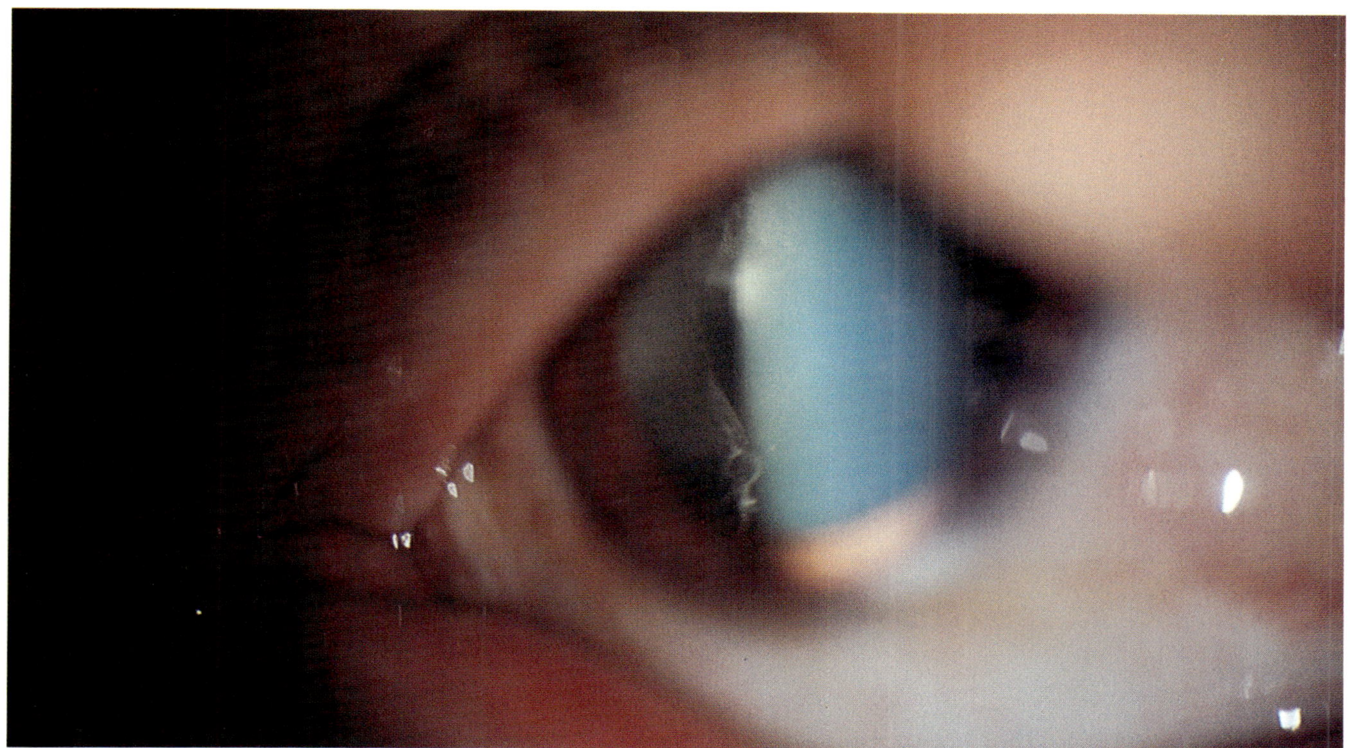

FIGURE 6–7 Clinical photograph shows yellow veils and strands within the vitreous body (amyloidosis). Courtesy of Dr. Kevin Miller, Jules Stein Eye Institute.

AMYLOIDOSIS

Vitreous amyloidosis is usually associated with primary familial amyloidosis. It is important to recognize vitreous amyloidosis because it may be the presenting sign or symptom of both systemic and ocular lesions, including proptosis, ocular palsies, internal ophthalmoplegia, neuroparalytic keratitis, glaucoma, and conjunctival involvement.

Amyloidosis in the vitreous is causally linked with retinal blood vessel involvement.[18–24] Vitreous amyloid may be dense enough to obscure vision or may be misinterpreted clinically as vitreous hemorrhage.[25] Biomicroscopically, the opacities may appear yellow-white and have a membranous veil-like quality (Figure 6–7). Some authors have described a central white dot associated with vertically oriented yellow strands.[26]

Vitrectomy may be done as a diagnostic or therapeutic maneuver.[27–30] Intraocular washings reveal amorphous birefringent deposits that are enhanced with Congo red stain (Figure 6–8). Red-green dichroism may be observed in specimens stained with Congo red and illuminated with polarized light (Figure 6–9). Amyloid fluoresces with thioflavin T stain. It has been demonstrated that vitreous amyloid reacts immunocytochemically with antibody to prealbumin.[31] Familial amyloidosis of the neuropathic type has been associated with homozygosity for the transthyretin methionine-30 gene.[32]

SILICONE VITREORETINOPATHY

Silicone oil is used in retinal detachment surgery as a means to tamponade retinal breaks and reattach the retina.[33–35] It is often reserved for cases of recurrent retinal detachments and advanced proliferative vitreoretinopathy.[36] There is experimental and clinical evidence to suggest that silicone oil may stimulate fibrous proliferation and recurrent membrane formation.[37–40] Histologic studies have demonstrated glial or pigment epithelial tissue containing vacuoles as well as extracellular clear spaces.[41] Cytologic preparations show intracellular vacuoles that presumably represent silicone oil engulfed by glial cells and macrophages (Figure 6–10).

HEALON-LADEN MACROPHAGES

The differential diagnosis of clear vacuoles within macrophages includes residual healon (hyaluronic acid polymer). This viscoelastic substance is used mainly during cataract surgery to keep the ocular chambers formed so instruments glide smoothly through incisions and the corneal endothelium is protected from damage. Cytologic specimens show macrophages with cytoplasmic clear spaces (Figure 6–11). A proliferative reaction has not been reported with healon.

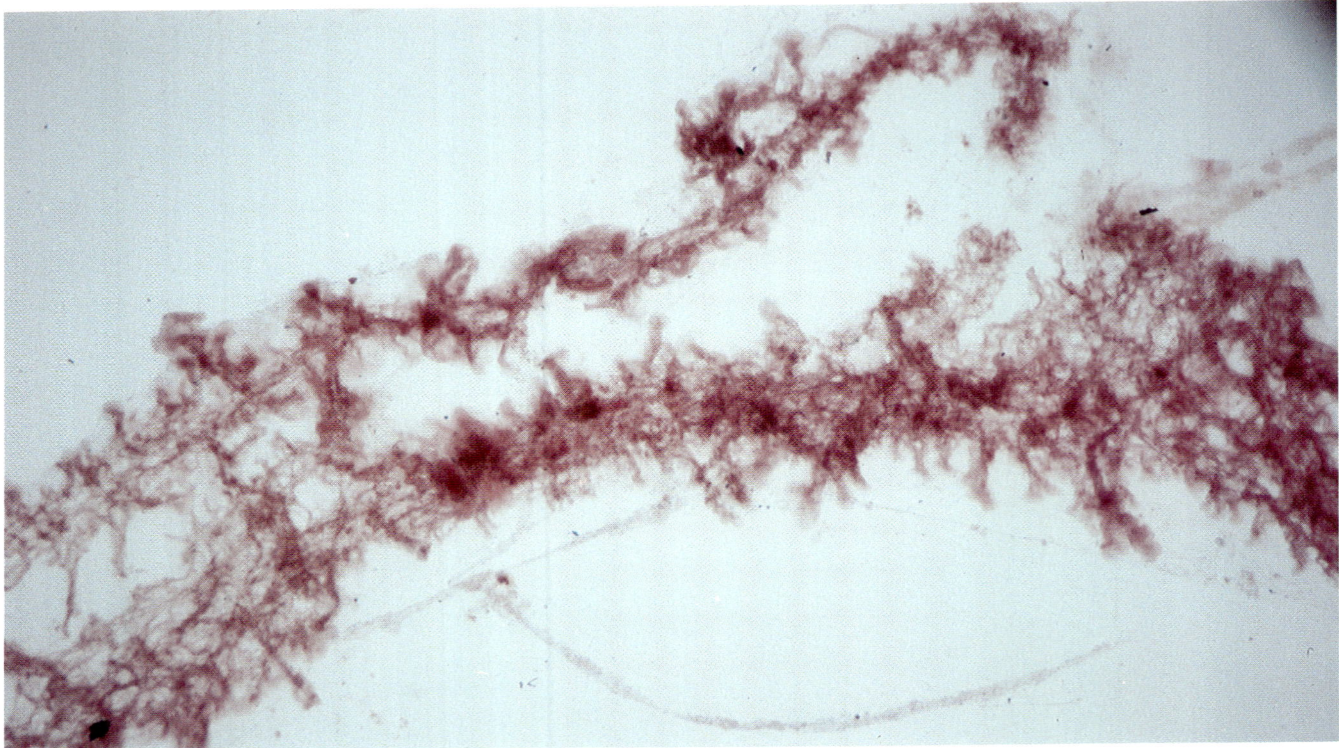

FIGURE 6–8 Photomicrograph from the intraocular washing from the patient shown in Figure 6–7 shows the globular formations of variable size (amyloid). (Congo red, × 250)

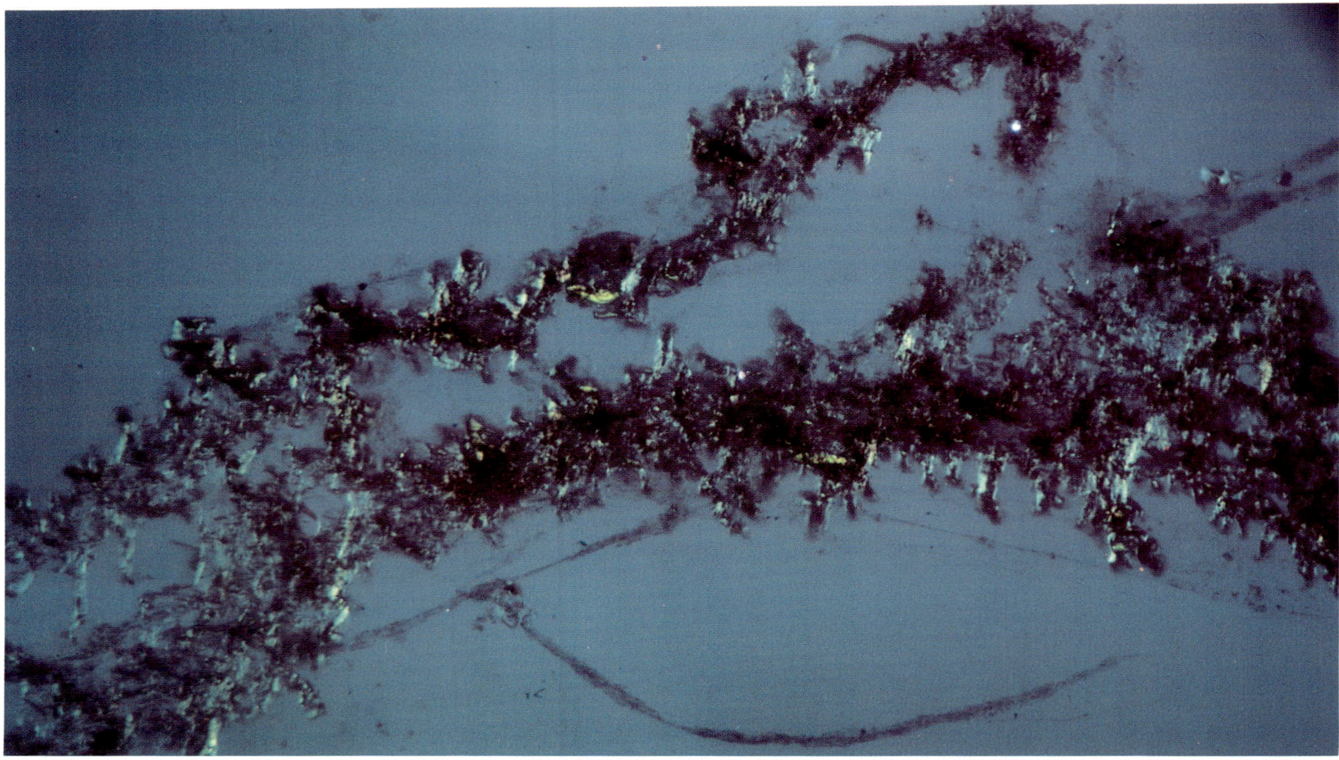

FIGURE 6–9 Congo red stains enhance the birefringence and dichroism seen in amyloid deposits in the vitreous. (Congo red and polarized light, × 250)

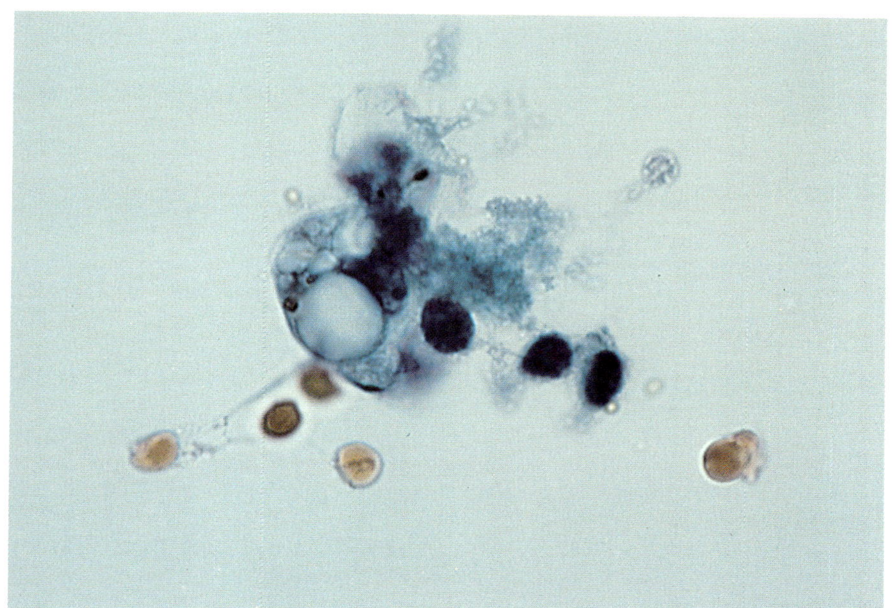

FIGURE 6–10 Photomicrograph of cells from an intraocular washing shows vacuoles, presumably displaced silicone oil. Surrounding glial cells and macrophages presumably proliferate in response to the silicone oil. (Papanicolaou, × 250)

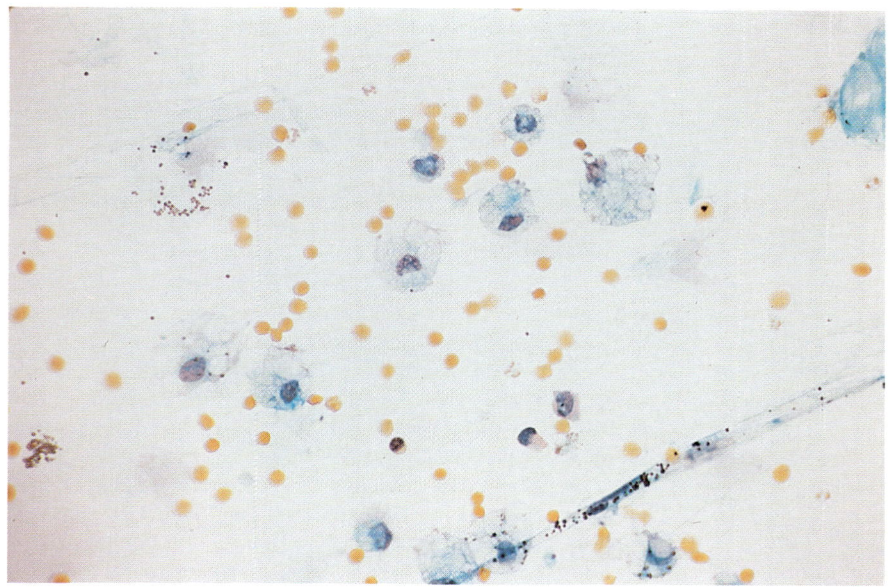

FIGURE 6–11 Photomicrograph shows macrophages with clear inclusions, residual healon. (Papanicolaou, × 125)

REFERENCES

1. Reese AB. Persistence and hyperplasia of the primary vitreous: retrolental fibroplasia-two entities. Arch Ophthalmol 1949;41:527–549.

2. Reese AB. Persistent hyperplastic primary vitreous. Am J Ophthalmol 1955;40:317–331.

3. Haddad R, Font RL, Reeser F. Persistent hyperplastic primary vitreous. A clinicopathologic study of 62 cases and review of the literature. Surv Ophthalmol 1978;23:123.

4. Font RL, Yanoff M, Zimmerman LE. Intraocular adipose tissue and persistent hyperplastic primary vitreous. Arch Ophthalmol 1969;82:43–50.

5. Manshot WA. Persistent hyperplastic primary vitreous. Arch Ophthalmol 1958;59:188–203.

6. Reese AB, Payne F. Persistence and hyperplasia of the primary vitreous. Am J Ophthalmol 1964;29:1–19.

7. Rutherford CW. Asteroid bodies in the vitreous. Arch Ophthalmol 1933;9:106.

8. Verhoeff FH. Microscopic findings in a case of asteroid hyalitis. Am J Ophthalmol 1921;4:155–160.

9. March W, Shock D, O'Grady R. Composition of asteroid bodies. Invest Ophthalmol 1974;13:701–705.

10. Miller H, Miller B, Rabinowitz H, et al. Asteroid bodies—an ultrastructural study. Invest Ophthalmol Vis Sci 1983;24:133–136.

11. Rodman HI, Johnson FB, Zimmerman LE. New histopathological and histochemical observations concerning asteroid hyalitis. Arch Ophthalmol 1961;66:552–563.

12. Engel HM, Green WR, Michels RG, Rice TA, Erozan YS, et al. Diagnostic vitrectomy. Retina 1981;1:121–149.

13. Renaldo DP. Pars plana vitrectomy for asteroid hyalosis. Retina 1981;1:252–254.

14. Lambrou FH, Sternberg P, Meredith TA, Mines J, Fine SL. Vitrectomy when asteroid hyalosis prevents laser photocoagulation. Ophthalmic Surg 1989;20:100–102.

15. Smith JL. Asteroid hyalitis: incidence of diabetes and hypercholesterolemia. JAMA 1958;168:891–893.

16. Hatfield AE, Gastineau CF, Rucker CW. Asteroid bodies in the vitreous: relationship to diabetes and hypercholesterolemia. Mayo Clin Proc 1962;37:513–514.

17. Luxenberg MN, Sime D. Relationship of asteroid hyalosis to diabetes mellitus and plasma lipid levels. Am J Ophthalmol 1969;67:406–413.

18. Lieberman TW, Ferry AP. Bilateral amyloidosis of the vitreous body without systemic involvement. Trans Am Ophthalmol Soc 1974;72:302–325.

19. Kaufman HE. Primary familial amyloidosis. Arch Ophthalmol 1958;60:1036–2043.

20. Crawford JB. Cotton wool exudates in systemic amyloidosis. Arch Ophthalmol 1967;78:214–216.

21. Wong VG, McFarlin DE. Primary familial amyloidosis. Arch Ophthalmol 1967;78:208–213.

22. Legrand J, Guenel J, Dubigeon P. Glaucome et opacification du vitre'par amylose. Bull Soc Ophtalmol Fr 1968;68:13–20.

23. Hamburg A. Unusual case of vitreous opacities. Primary familial amyloidosis. Ophthalmologica 1971; 162:173–177.

24. Franceschetti AT, Rabinowicz T. Les lesions oculaires dans l'amyloidose primaire familiale. J Genet Hum 1969;17:349–366.

25. Schwartz MF, Green WR, Michels RG, Kincaid MC, Fogle J. An unusual case of ocular involvement in primary systemic nonfamilial amyloidosis. Ophthalmology 1982; 89:394–401.

26. Mieler WF, Williams DF, Levin M. Vitreous amyloidosis. Arch Ophthalmol 1988;106:881–883.

27. Paton D, Duke JR. Primary familial amyloidosis. Ocular manifestations with histopathologic observations. Am J Ophthalmol 1966;51:736–747.

28. Kasner D, Miller GR, Taylor WH, Sever RJ, Norton EW, et al. Surgical treatment of amyloidosis of the vitreous. Trans Am Acad Ophthalmol Otolaryngol 1968;72:410–418.

29. Matsui M, Tashiro T, Asai Y. A case of bilateral vitreous amyloidosis which was successfully treated with vitreous surgery. Acta Soc Ophthalmol Jpn 1976;80:142–146.

30. Irvine AR, Char DH. Recurrent amyloid involvement in the vitreous body after vitrectomy. Am J Ophthalmol 1976;82:705–708.

31. Doft BH, Rubinow A, Cohen AS. Immunocytochemical demonstration of prealbumin in the vitreous in heredofamilial amyloidosis. Am J Ophthalmol 1984;97:296–300.

32. Sandgren O, Holgrem G, Lundgren E. Vitreous amyloidosis associated with homozygosity for the transthyretin methionine-30 gene. Arch Ophthalmol 1990;108:1584–1586.

33. Cibis PA, Becker B, Okun E, Canaan S. The use of silicone in retinal detachment surgery. Arch Ophthalmol 1962;68:590–599.

34. Grey RHB, Leaver PK. Silicone oil in the treatment of massive preretinal traction. 1. Results of 105 eyes. BR J Ophthalmol 1979;63:355–360.

35. McCuen BW II, Landers MB III, Machemer R. The use of silicone oil following failed vitrectomy for retinal detachment with advanced proliferative vitreoretinopathy. Ophthalmology 1985;92:1029–1034.

36. Cos MS, Trese MT, Murphy PL. Silicone oil for advanced proliferative vitreoretinopathy. Ophthalmology 1986;93:646–650.

37. Fastenberg DM, Diddie KR, Delmage JM, Dorey K. Intraocular injection of silicone oil for experimental proliferative vitreoretinopathy. Am J Ophthalmol 1983; 95:663–667.

38. Gonvers M, Thresher R. Temporary use of silicone oil in the treatment of proliferative vitreoretinopathy: an experimental study with a new animal model. Graefes Arch Clin Exp Ophthalmol 1983;221:46–53.

39. Watzke RC. Silicone retinopiesis for retinal detachment: a long-term clinical evaluation. Arch Ophthalmol 1967;77:185–196.

40. Cockerham WD, Schepens CL, Freeman HM. Silicone injection in retinal detachment. Arch Ophthalmol 1970;83:704–712.

41. Lewis H, Burke JM, Abrams GW, Aaberg TM. Perisilicone proliferation after vitrectomy for proliferative vitreoretinopathy. Ophthalmology 1988;95:583–591.

CHAPTER

7 Abnormalities of the Retina

Our goal in this chapter is to introduce those retinal diseases that are most commonly encountered in cytology specimens. The diagnosis of many of these lesions by cytologic techniques alone can be difficult and thorough familiarity with the tissue lesion can be very helpful.

The ocular cytologist is likely to encounter specimens from only a few retinal diseases on a routine basis. Intraocular washings from patients with diabetic retinopathy and proliferative vitreoretinopathy account for most of the specimens. Because the diagnosis is often apparent from clinical examination, some argue that it is superfluous to send the intraocular washings for pathologic examination. Indeed, only 15% of intraocular washings are diagnostic of specific diseases.[1] However, the need to exclude clinically unsuspected infectious and neoplastic disorders justifies routine processing of vitrectomy specimens. Examination of intraocular washings sometimes reveals retinal fragments. It is often difficult to diagnose specific diseases from retinal fragments seen in cytologic preparations. Other accompanying fragments usually provide clues to the diagnosis. Notable exceptions in which diseases are diagnosed by retinal biopsies include infectious retinitis and retinoblastoma. These entities are discussed separately in Chapters 8 and 9, respectively.

RETINOPATHY OF PREMATURITY

Previously called retrolental fibroplasia,[2] retinopathy of prematurity is a disease of the premature infant.[3] It is usually, but not always, associated with exposure to high concentrations of oxygen.[4–7] The natural history and pathogenesis of this disease is discussed in detail elsewhere.[8–11] In brief, retinopathy of prematurity develops at the border of the developing retinal vasculature.[12,13] Presumably, ischemia leads to proliferation of the primitive microvasculature, which in turn may lead to tractional retinal detachment (Figure 7–1).[14,15]

Vitrectomies are performed in the late stages to remove the vitreous and fibrovascular tissue in order to repair the retinal detachment.[16,17] When the retina is detached, the visual prognosis is poor.[18] In cytologic preparations, fibrous tissue, lens fragments (lensectomy), pigment-laden macrophages, and vitreous hemorrhage can be observed (Figure 7–2). These findings are not specific for retinopathy of prematurity.

COATS' DISEASE

Coats' disease is usually unilateral and is characterized by exudative retinal detachment in young males.[19–21] Telangiectatic blood vessels have been identified in this disorder and presumably are the basis for the retinal exudates and exudative detachment of the retina (Figure 7–3).[22,23] Coats' disease may produce leukocoria in children and can be confused clinically with retinoblastoma. In such cases, a fine needle aspiration may be obtained to differentiate these two entities. In addition, a vitrectomy may be employed to repair the retinal detachment.[24–26] In either case, the specimens are extremely hypocellular with occasional macrophages containing cytoplasmic vacuoles (Figure 7–4).[27] Histologically, there is massive subretinal exudate and retinal detachment. Telangiectasis of the retinal blood vessels or the extraretinal blood vessels can be identified (Figure 7–5).

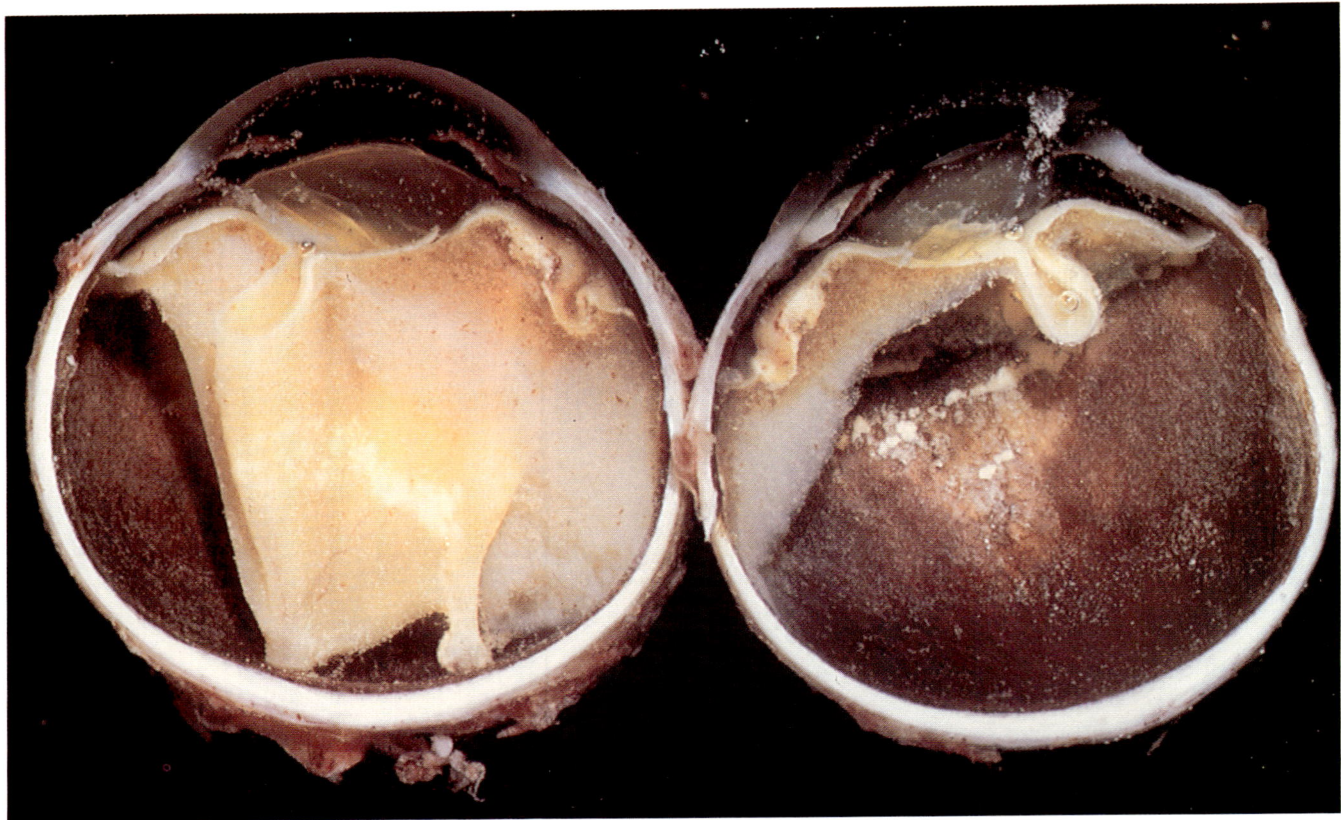

FIGURE 7–1 Gross photograph of a tractional retinal detachment in retinopathy of prematurity. Published courtesy of Ophthalmology 1985; 92:563–574.

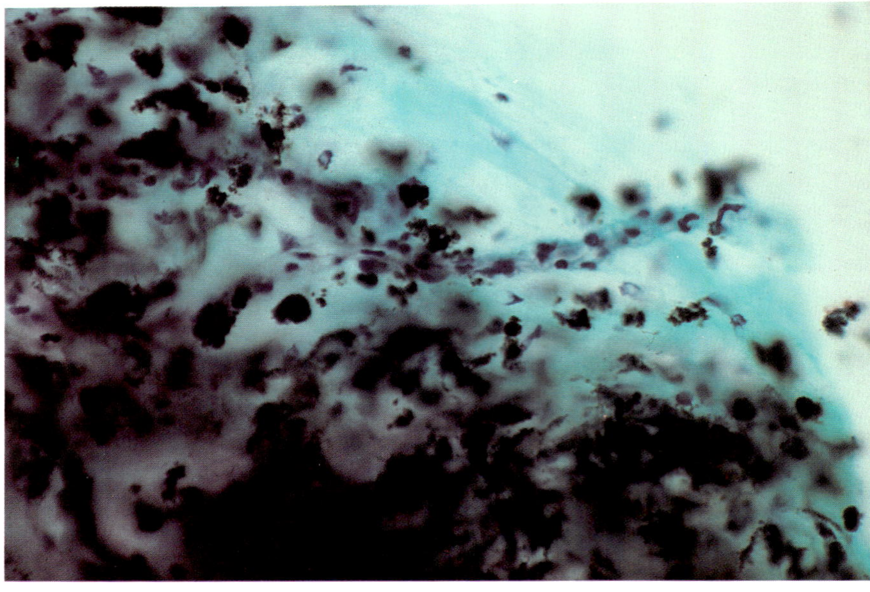

FIGURE 7–2 Photomicrograph of cytospin preparation from a patient with a tractional detachment from retinopathy of prematurity. A small blood vessel (neovascularization) courses through numerous spindle cells and pigment-laden macrophages. (Papanicolaou, × 250)

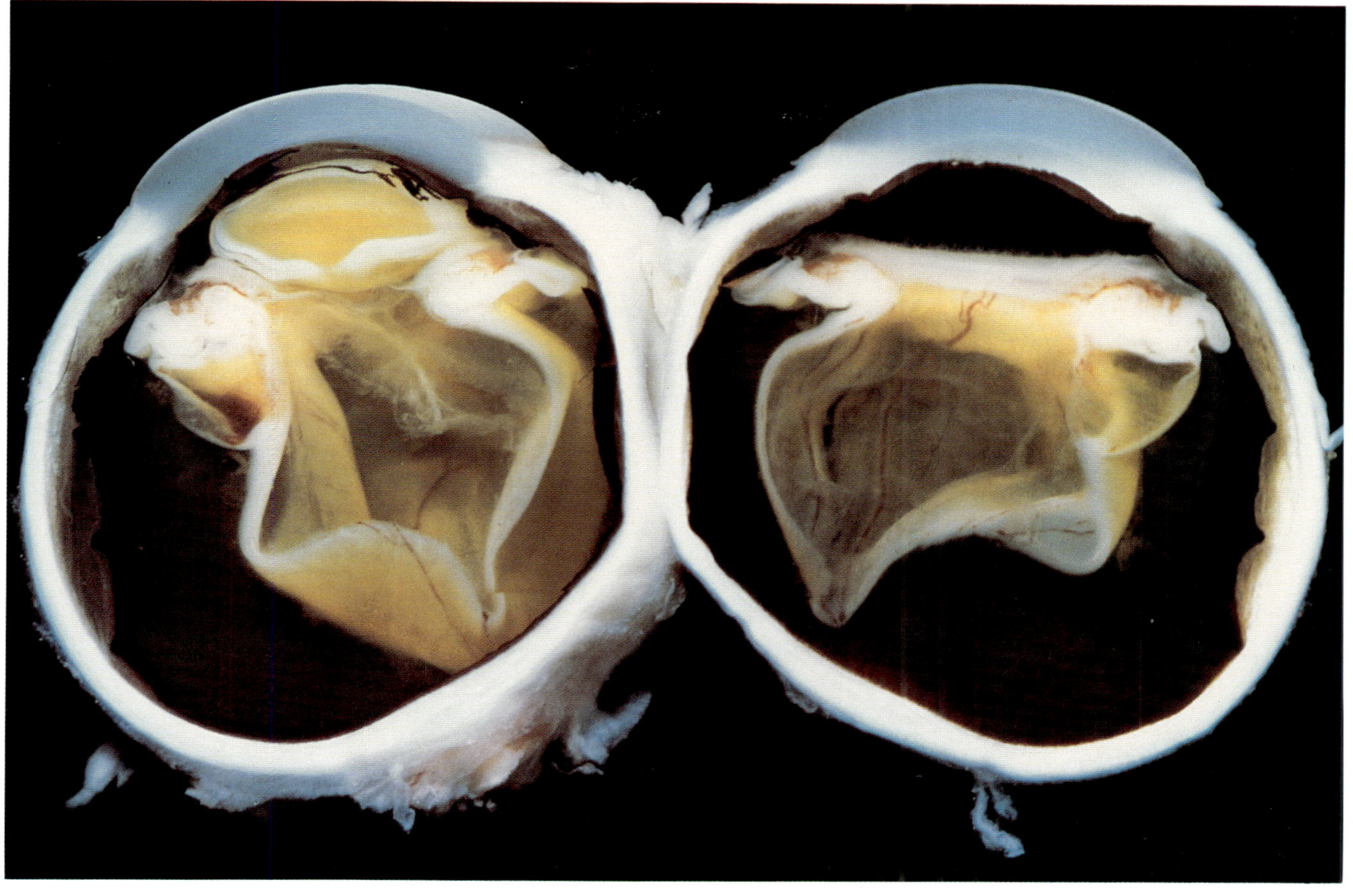

FIGURE 7–3 Gross photograph of a patient with Coats' disease shows extensive exudative retinal detachment and dilated blood vessels.

DIABETIC RETINOPATHY

Diabetic retinopathy is the single most common source of intraocular washing specimens processed by our cytology laboratory. Retinopathy occurs in about 70% of patients who have had diabetes mellitus beyond 20 years. Diabetic retinopathy manifests as a microangiopathy with thickening of vascular basement membrane, loss of endothelium, and degeneration of capillary pericytes.[28–32] In the early stages, venous dilatation, capillary microaneurysms, and exudates are observed.[33,34] Vascular disease leads to ischemia of the retina and neovascularization is stimulated (proliferative diabetic retinopathy) (Figure 7–6).[35,36] Vitrectomy may be performed in this disease for several reasons, including vitreous hemorrhage secondary to neovascularization and tractional retinal detachment with removal of fibrovascular membranes.[37–39]

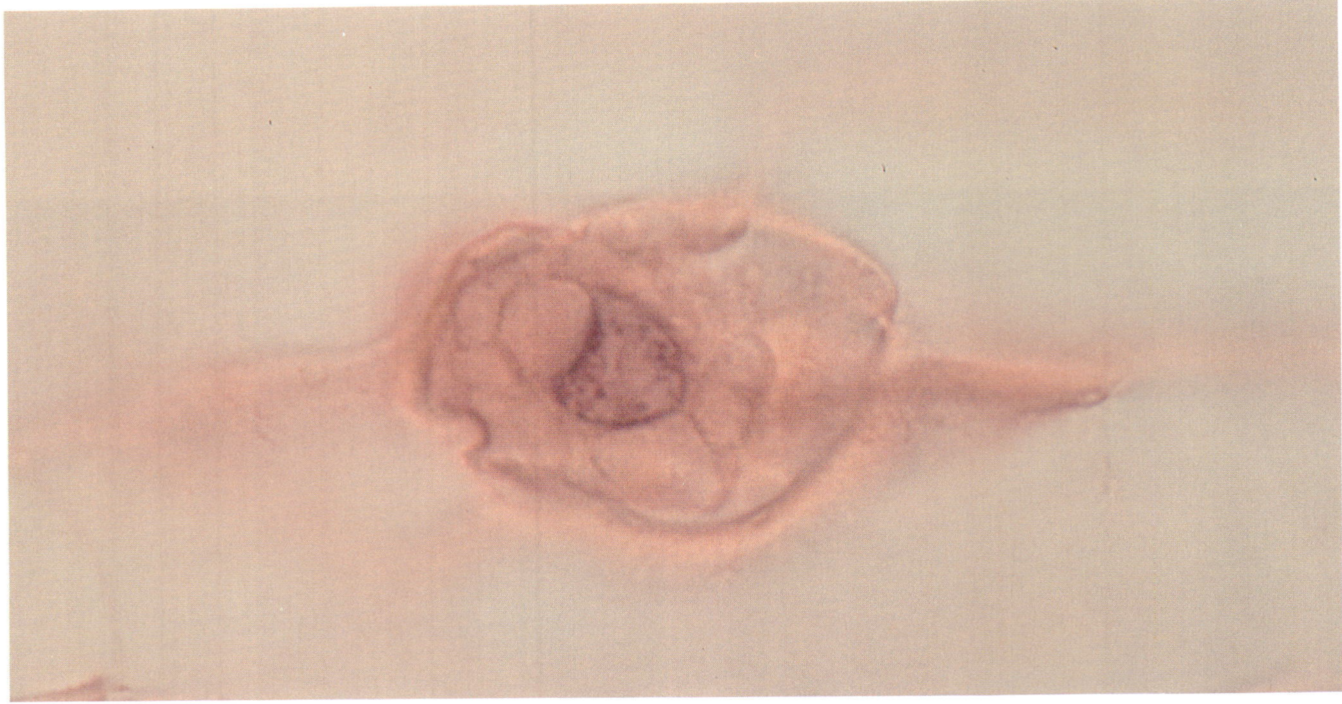

FIGURE 7–4 (ABOVE) Photomicrograph of a fine needle aspiration from a patient with Coats' disease. Macrophages with a glossy surface are covered and embedded in a thick exudate. Cytoplasmic vacuoles (presumably engulfed lipid) displace the nucleus. (hematoxylin and eosin, × 1000)

FIGURE 7–5 (BELOW) Histologic section from the eye of a patient who clinically was diagnosed with possible Coats' disease. The eye was enucleated because retinoblastoma could not be excluded. Massive proteinaceous exudate lies beneath a totally detached retina. (hematoxylin and eosin, × 10)

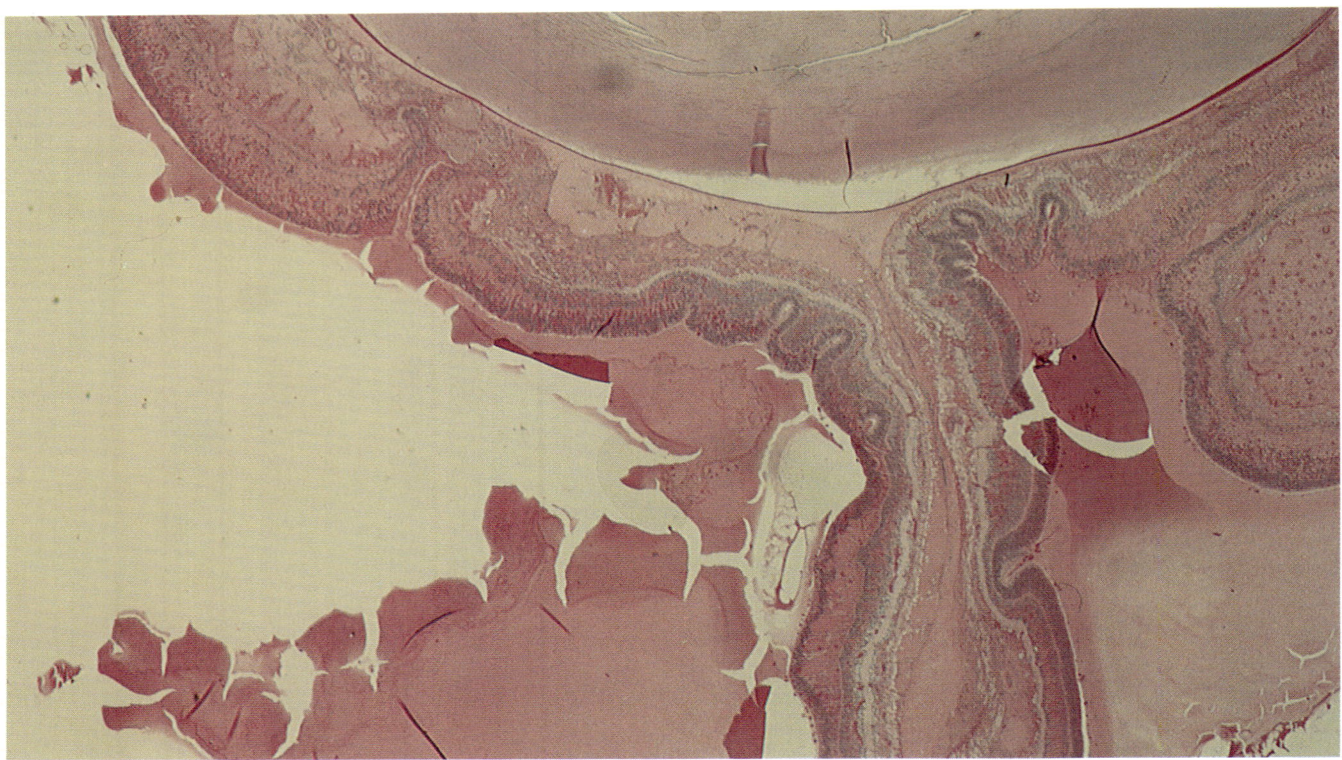

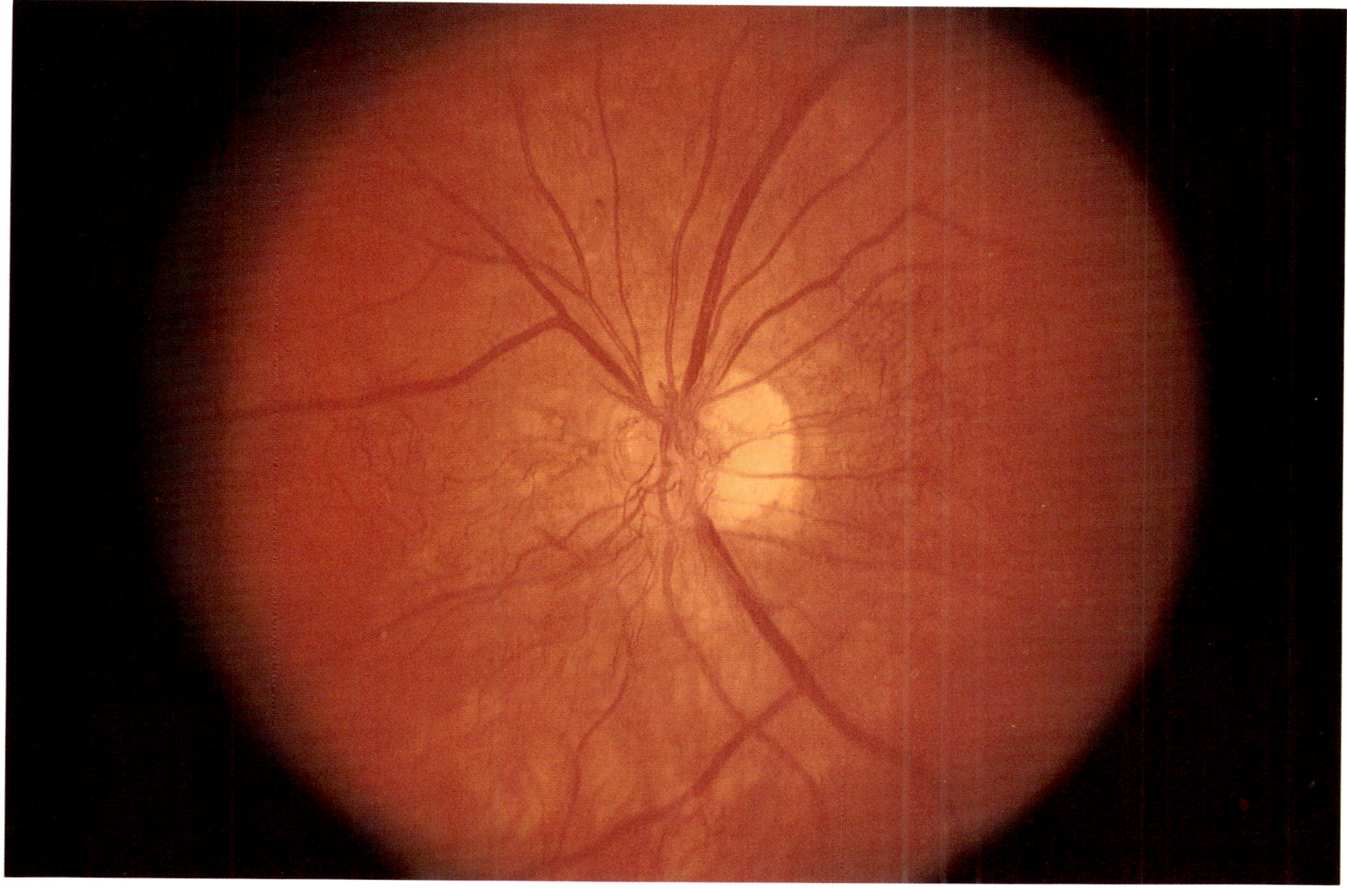

FIGURE 7–6 **Fundus photograph of a diabetic patient shows extraretinal blood vessels emanating from the optic nerve head (proliferative diabetic retinopathy).**

Neovascularization is identified in vitrectomy specimens as well-defined blood vessels lined by a single layer of endothelium (Figure 7–7).

PROLIFERATIVE VITREORETINOPATHY

The leading cause of failure in retinal detachment surgery is proliferative vitreoretinopathy (PVR).[40–43] It is characterized by a proliferation of cellular membranes on both sides of the detached retina and in the vitreous cavity.[44] Fibrovascular proliferation that extends anterior to the posterior border of the vitreous base has been termed anterior PVR.[45] Traction induced from the membranes on the posterior lens surface, ciliary body, and iris may produce a peripheral retinal detachment.[46] Epiretinal membranes are composed of retinal pigment epithelial cells, glial cells, fibrocytes, and macrophages.[47–56] Surgically removed epiretinal membranes may be sent for cytologic examination. Cell buttons or cytospin preparations reveal a thin fragment composed of spindle cells and oval cells, some of which may contain pigment (Figure 7–8).

Epiretinal membranes composed mainly of spindle cells have been associated with photocoagulation, trauma, and inflammatory disease.[57] However, most simple epiretinal membranes are idiopathic.[58]

EALES' DISEASE

Eales' disease was initially described in young males with vitreous hemorrhage and abnormal retinal veins.[59] Now it is clinically defined by an obliterative vasculopathy involving both venules and arterioles (Figure 7–9).[60] In the late stages, ischemia leads to neovascularization with resulting vitreous hemorrhage, retinal detachment, and secondary glaucoma.[61,62] Specimens are obtained when vitrectomy is done to remove persistent vitreous hemorrhage or extraretinal fibrous tissue.[63] Cytologic specimens may show vitreous hemorrhage (hemoglobin spherulosis) and, occasionally, fibrous tissue (Figure 7–10).

Other conditions that produce neovascularization and vitreous hemorrhage include branch retinal vein occlusion,[64] sickle cell retinopathy,[65] carotid and aortic arch occlusive syndromes,[66] and hyperviscosity syndromes.[67]

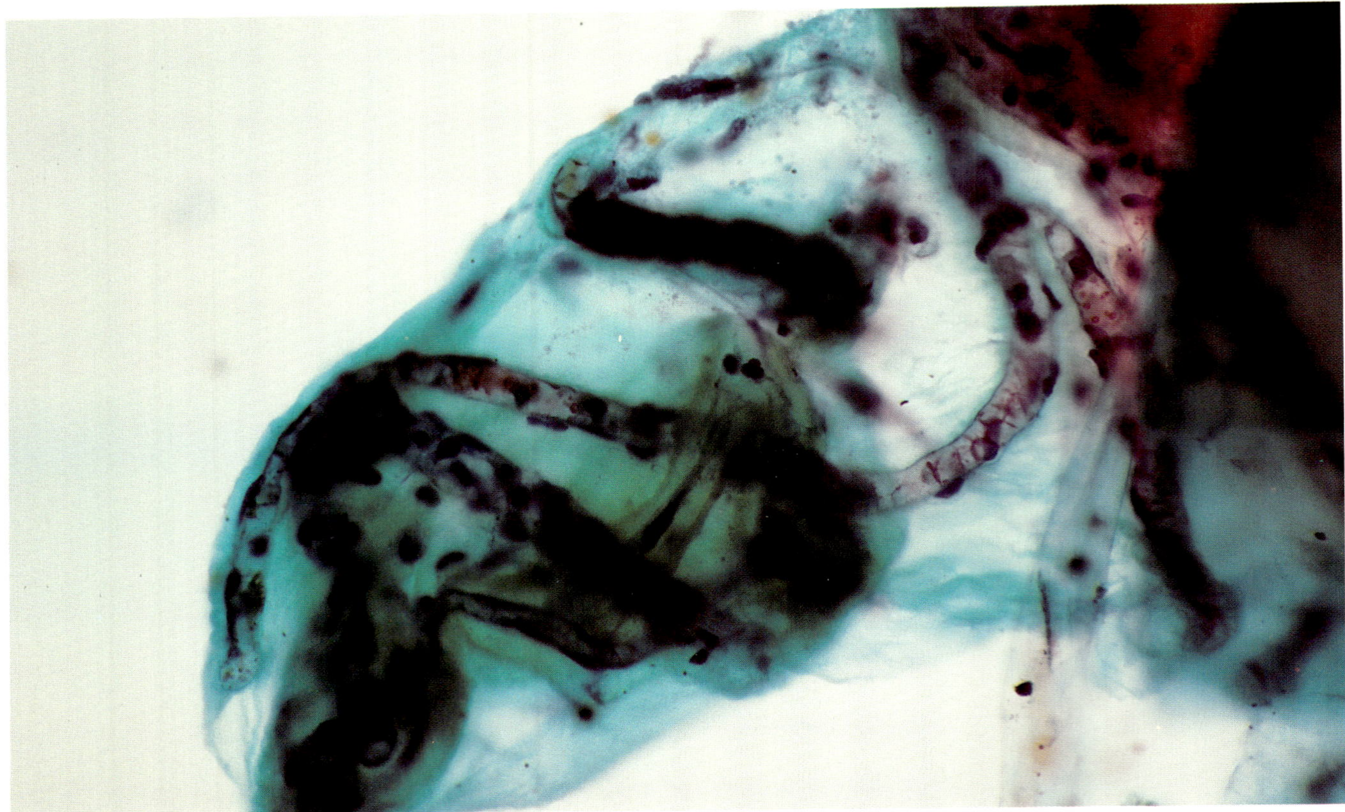

FIGURE 7–7 (ABOVE) Photomicrograph of intraocular washing specimen from a patient with proliferative diabetic retinopathy. Branching blood vessels containing blood cells and lined by single endothelial cells are identified. (Papanicolaou, × 250)

FIGURE 7–8 (BELOW) Photomicrograph of an epiretinal membrane containing spindle and oval cells with pigment. The wavy membrane is the internal limiting lamina of the retina. (periodic acid-Schiff, × 125)

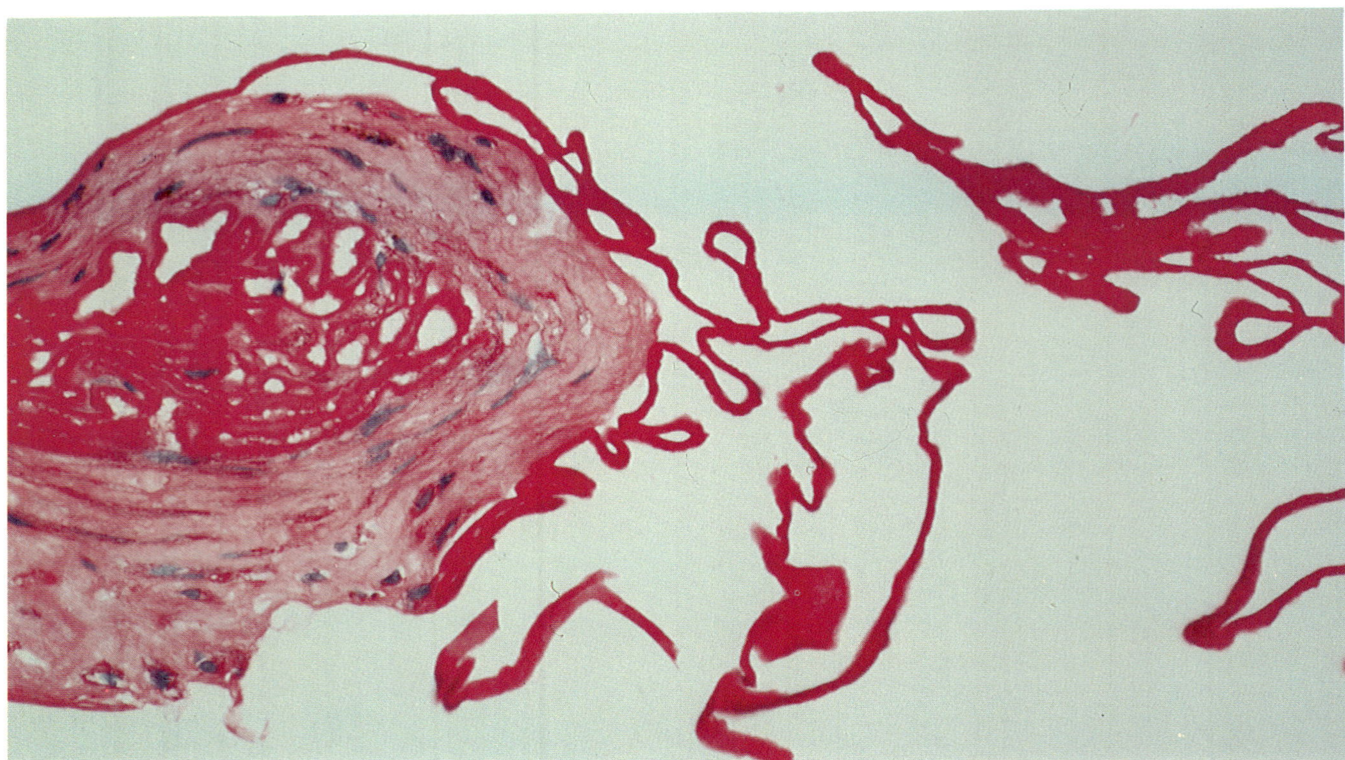

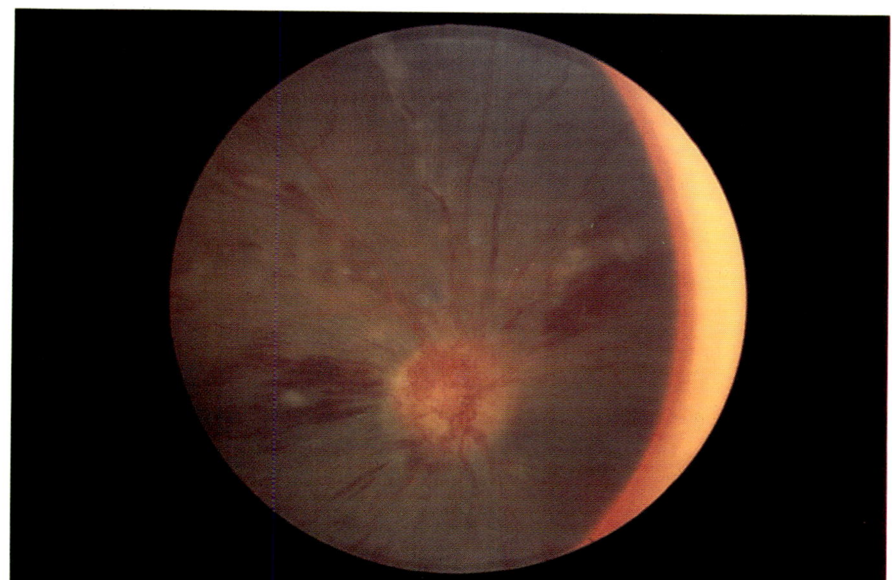

FIGURE 7–9 Clinical photograph shows the abnormal perivascular sheathing and extraretinal neovascularization in a patient clinically diagnosed to have Eales' disease.

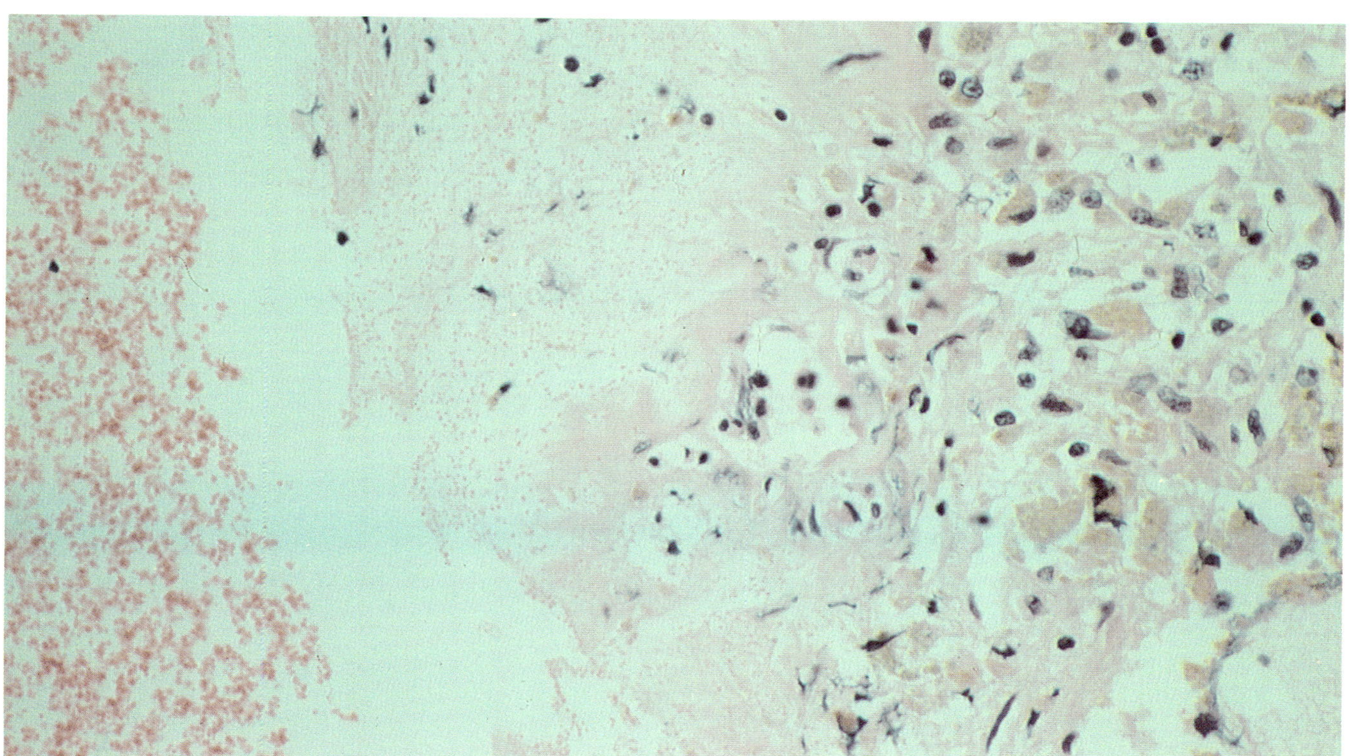

FIGURE 7–10 Photomicrograph from the cytospin preparation of intraocular washings that were obtained from the patient shown in Figure 7–9. Spindle cells, hemosiderin-laden macrophages, and inflammatory cells (left) are juxtaposed next to numerous variably sized red globules (hemoglobin spherulosis) indicative of prior vitreous hemorrhage. (hematoxylin and eosin, × 250)

REFERENCES

1. Green WR. Diagnostic cytopathology of ocular fluid specimens. Ophthalmology 1984;91:726–749.

2. Foos RY. Acute retrolental fibroplasia. Albrecht Von Graefes Arch Klin Exp Ophthalmol 1975;195:87–100.

3. Keith CG, Kitchen WH. Retinopathy of prematurity in extremely low birthweight infants. Med J Aust 1984;141:225–227.

4. Patz A. Observations on the retinopathy of prematurity. Am J Ophthalmol 1985;100:164–168.

5. Kushner BJ, Gloeckner E. Retrolental fibroplasia in full-term infants without exposure to supplemental oxygen. Am J Ophthalmol 1984;27:148–163.

6. Shohat M, Reisner SH, Krikler R, Nissonkorn I, Yassur Y, et al. Retinopathy of prematurity. Incidence and risk factors. Pediatrics 1983;72:159–163.

7. Karlsberg RC, Green WR, Patz A. Congenital retrolental fibroplasia. Arch Ophthalmol 1973;89:122–123.

8. Flynn JT, Bancalari E, Bachynski BN, Buckley EB, Bawd R, et al. Retinopathy of prematurity. Diagnosis, severity and natural history. Ophthalmology 1987;94:620–629.

9. Prendiville A, Schulenberg WE. Clinical factors associated with retinopathy of prematurity. Arch Dis Child 1988; 63:522–527.

10. Silverman WA, Flynn JT. Retinopathy of prematurity. Boston: Blackwell Scientific, 1985.

11. McPherson AR, Hittner HM, Kretzer FL. Retinopathy of prematurity. Current concepts and controversies. Toronto: B.C. Decker, 1986.

12. Foos RY. Retinopathy of prematurity. Pathologic correlation of clinical stages. Retina 1987;7:260–276.

13. Kushner BJ, Essner D, Cohen IJ, Flynn JT. Retrolental fibroplasia II. Pathologic correlation. Arch Ophthalmol 1977; 95:29–38.

14. Campbell FW. The influence of a low atmospheric pressure on the development of the retinal vessels in the rat. Trans Ophthalmol Soc UK 1951;71:287–300.

15. Quinn GE, Schaffer DB, Johnson L. A revised classification of retinopathy of prematurity. Am J Ophthalmol 1982;94:744–749.

16. Lightfoot D, Irvine AR, Vitrectomy in infants and children with retinal detachments caused by cicatricial retrolental fibroplasia. Am J Ophthalmol 1982;94:305–312.

17. Schepens CL. Clinical and research aspects of subtotal open-sky vitrectomy. Am J Ophthalmol 1981;91:143–171.

18. Jabour NM, Eller AE, Hirose T, Schepens CL, Liberfarb R, et al. Stage 5 retinopathy of prematurity. Prognostic value of morphologic findings. Ophthalmology 1987;94:1640.

19. Coats G. Forms of retinal disease with massive exudation. Royal London Ophthal Hosp Rep 1908;17:440–525.

20. Fox KR. Coats' disease. Metabol Pediatr Syst Ophthalmol 1980;4:121–124.

21. Reese AB. Telangiectasis of the retina and Coats' disease. Am J Ophthalmol 1956;42:1–8.

22. Chang M, McLean IW, Merritt JC. Coats' disease. A study of 62 histologically confirmed cases. J Pediatr Ophthalmol Strabismus 1984;21:163–168.

23. Egbert PR, Chan CC, Winter FC. Flat preparations of the retinal vessels in Coats' disease. J Pediatr Ophthalmol 1977;13:336–339.

24. Jaffe MS, Shields JA, Canny CLB, Eagle RC, Fry RL. Retinoblastoma simulating Coats' disease: a clinicopathologic report. Ann Ophthalmol 1977;9:863–868.

25. Pe'er J. Calcification in Coats' disease. Am J Ophthalmol 1988;106:742–743.

26. Laqua H, Wessing A. Peripheral retinal telangiectasis in adults simulating a vascular tumor or melanoma. Ophthalmology 1983;90:1284–1291.

27. Haik BG, Koizumi J, Smith ME, Ellsworth, RM. Fresh preparation of subretinal fluid aspirations in Coats' disease. Am J Ophthalmol 1985;100:327–328.

28. Robison WG Jr, Kador PF, Kinoshita JH. Retinal capillaries: basement membrane thickening by galactosemia prevented with aldose reductase inhibitor. Science 1983; 221:1177.

29. Kalebic T, Garbisa S, Glaser B, Liotta LA. Basement membrane collagen: degradation by migrating endothelial cells. Science 1983;221:281.

30. Kuwabara T, Cogan DG. Retinal vascular patterns. VI. Mural cells of the retinal capillaries. Arch Ophthalmol 1963;69:492–502.

31. Toussaint D, Dustin P. Electron microscopy of normal and diabetic retinal capillaries. Arch Ophthalmol 1963;70:96–108.

32. Yanoff M. Diabetic retinopathy. N Engl J Med 1966;274:1344–1349.

33. De Venecia G, Davis M, Engerman R. Clinicopathologic correlations in diabetic retinopathy. I. Histology and fluorescein angiography of microaneurysms. Arch Ophthalmol 1976;94:1766–1773.

34. Addison DJ, Garner A, Ashton N. Degeneration of intramural pericytes in diabetic retinopathy. Br Med J 1970;1: 264–266.

35. Niki T, Muraoka K, Shimizu K. Distribution of capillary nonperfusion in early-stage diabetic retinopathy. Ophthalmology 1984;91:1440.

36. Miller H, Miller B, Zonis S, Nir I. Diabetic neovascularization: permeability and ultrastructure. Invest Ophthalmol Vis Sci 1984;25:1338–1342.

37. Thompson JT, de Bustros S, Michels RG, Rice TA. Results and prognostic factors in vitrectomy for diabetic traction-rhegmatogenous retinal detachment. Arch Ophthalmol 1987;105:503–507.

38. Thompson JT, de Bustros S, Michels RG, Rice TA. Results and prognostic factors in vitrectomy for diabetic vitreous hemorrhage. Arch Ophthalmol 1987;105:191–195.

39. Blankenship GW. Proliferative diabetic retinopathy: principles and techniques of surgical treatment. In: Ryan SJ, Glaser BM, Michels RG, eds. Retina. St Louis: C.V. Mosby, 1989;3:515–539.

40. The Retinal Society Terminology Committee. The classification of retinal detachment with proliferative vitreoretinopathy. Ophthalmology 1983;90:121–125.

41. Claes C, Freeman HM, Tolentino FI. PVR: An overview. In: Freeman HM, Tolentino FI, eds. Proliferative vitreoretinopathy (PVR). New York: Springer-Verlag, 1988:3–11.

42. Lewis H, Aaberg TM, Abrams GW. Causes of failure after initial vitreoretinal surgery for severe proliferative vitreoretinopathy. Am J Ophthalmol 1991;111:8–14.

43. Schwartz D, De la Cruz ZC, Green WR, Michels RG. Proliferative vitreoretinopathy. Ultrastructural study of 20 retroretinal membranes removed by vitreous surgery. Retina 1988;8:275–281.

44. Glaser BM. Pathobiology of PVR. In: Freeman HM, Tolentino FI, eds. Proliferative vitreoretinopathy (PVR). New York: Springer-Verlag, 1988:12–21.

45. Lewis H, Aaberg T. Anterior proliferative vitreoretinopathy. Am J Ophthalmol 1988;105:277.

46. Lewis H, Abrams GW, Foos RY. Clinicopathologic findings in anterior hyaloidal fibrovascular proliferation after diabetic vitrectomy. Am J Ophthalmol 1987;104:614–618.

47. Machemer R, Laqua H. Pigment epithelium proliferation in retinal detachment (massive periretinal proliferation). Am J Ophthalmol 1975;80:1–23.

48. Clarkson JG, Green WR, Massoff D. A histopathologic review of 168 cases of preretinal membrane. Am J Ophthalmol 1977;84:1–17.

49. Daicker B, Guggenheim R. Studies on pigmented surface wrinkling retinopathy by scanning electron microscopy. Albrecht Von Graefes Arch Klin Exp Ophthalmol 1979; 210:109–110.

50. Van Horn DL, Aaberg TM, Machemer R. Glial cell proliferation in human retinal detachment with massive periretinal proliferation. Am J Ophthalmol 1977;84:383–393.

51. Kampik A, Kenyon KR, Michels RG, Green WR, de la Cruz ZC. Epiretinal and vitreous membranes: comparative study of 56 cases. Arch Ophthalmol 1981;99:1445–1454.

52. Kampik A, Green WR, Michels RG, Nase PK. Ultrastructure of progressive idiopathic epiretinal membrane removed by vitreous surgery. Am J Ophthalmol 1980;90: 797–809.

53. Hiscott PS, Grierson I, McLeod D. Natural history of fibrocellular epiretinal membranes: a quantitative, autoradiographic and immunohistochemical study. Br J Ophthalmol 1985;69:810–823.

54. Michels RG. A clinical and histopathological study of epiretinal membranes affecting the macula and removed by vitreous surgery. Trans Am Ophthalmol Soc 1982;80:580–656.

55. Foos RY. Spectrum of nonvascular proliferative extraretinopathies. In: Nicholson DH, ed. Ocular pathology update. New York: Masson Publishing, 1980:107–114.

56. Foos RY. Nonvascular proliferative extraretinopathies. Am J Ophthalmol 1978;86:723–725.

57. Kampik A, Green WR, Michels RG, Rice TA. Epiretinale membranen nack photokoagulation (postkoagulative maculopathie). Ber Dtsch Ophthalmol Ges 1981; 78:593–598.

58. Roth AM, Foos RY. Surface wrinkling retinopathy in eyes enucleated at autopsy. Trans Am Acad Ophthalmol Otolaryngol 1971;75:1047–1058.

59. Eales H. Cases of retinal hemorrhage associated with epistaxis and constipation. Birm Med Rev 1880;9:262.

60. Elliot AJ. Thirty-year observation of patients with Eales' disease. Am J Ophthalmol 1975;80:404–408.

61. Renie WA, Murphy RP, Anderson KC, Lippman SM, McKusick VA, et al. The evaluation of patients with Eales' disease. Retina 1983;3:243–248.

62. Spitznas M, Meyer-Schwickerath GT, Staphan B. The clinical picture of Eales' disease. Albrecht Von Graefes Arch Klin Exp Ophthalmol 1975;194:73–85.

63. Gieser SC, Murphy RP. Eales' disease. In: Ryan SJ, Schachat AP, Murphy RP, Patz A, eds. Retina. St Louis: C.V. Mosby, 1989;2:535–540.

64. Hayreh SS, Rojas P, Podhajsky P, Montague P, Woolson RF. Ocular neovascularization with retinal vascular occlusion. III. Incidence of ocular neovascularization with retinal vein occlusion. Ophthalmology 1983;20:488–506.

65. Goldberg MF. Classification and pathogenesis of proliferative sickle retinopathy. Am J Ophthalmol 1971;71: 649–665.

66. Kahn M, Green WR, Knox DL, Miller NR. Ocular features of carotid occlusive disease. Retina 1986;6:239–252.

67. Ring CP, Pearson TC, Sanders MD, Wetherly-Mein G. Viscosity and retinal vein thrombosis. Br J Ophthalmol 1967;60:397–410.

CHAPTER 8

Intraocular Infections

In this chapter, selected infectious diseases that have significance in ocular cytology are presented. External infections such as *Chlamydia trachomatis* and adenovirus were discussed in Chapter 3. Endophthalmitis may be divided clinically into endogenous (arising from systemic infection) and exogenous (arising from external sources).

BACTERIAL ENDOPHTHALMITIS

Over 90% of cases of endophthalmitis originate from bacterial sources and the majority of these are caused by gram-positive organisms (*Staphylococcus* sp., most commonly).[1]

Gram-negative organisms that are associated with endophthalmitis include *Proteus, Klebsiella* sp., *Serratia marcescens, Haemophilus* spp., and *Pseudomonas aeruginosa.* In endophthalmitis associated with cataract surgery, *S. epidermidis* is the most commonly identified organism.[2,3] Other organisms that frequently cause infection include *S. aureus* and *Propionibacterium acnes.*[4] In post-traumatic endophthalmitis, *Bacillus* spp. are second only to *staphylococcus* in incidence. After glaucoma filtering procedures, *Streptococcus* is most common.[5] Endogenous endophthalmitis is associated with intravenous drug abuse, meningitis, endocarditis, and urinary tract infections.[6] The most common organisms isolated in endogenous endophthalmitis include *Bacillus* sp., *streptococcus* sp., *Neisseria meningitidis, S. aureus*, and *H. influenzae.*

Cytology specimens are obtained in cases of acute endophthalmitis for culture and morphologic diagnosis. If a diagnostic tap of the aqueous or vitreous is performed, cultures, gram stain, and cytology should be requested (Figure 8–1).

In acute endophthalmitis, direct smears show numerous neutrophils and fibrin. Bacteria are best revealed on gram stain. Frequently, the aqueous aspirate is negative and a vitreous aspirate is necessary to demonstrate organisms (Figures 8–2 and 8–3).

Vitrectomy may be indicated for both diagnosis and therapeutic removal of bacteria, fibrin, and necrotic material.[7] Cytology preparations will reveal abundant acute inflammatory cells and necrotic debris (Figure 8–4). Normal structures may be removed inadvertently in a vitrectomy because intraoperative identification and separation of inflamed tissues is difficult (Figures 8–5 and 8–6).

MYCOBACTERIUM FORTUITUM

Mycobacterium fortuitum infection in the eye usually manifests as a suppurative keratitis related to trauma, including contact lens use, or an operation.[8–12] *M. fortuitum* rarely infects the vitreous cavity primarily. However, corneal infection may spread to involve the vitreous. Intraocular washings demonstrate numerous intracellular acid-fast long-curving bacilli (Figure 8–7).

FUNGAL ENDOPHTHALMITIS

Fungal endophthalmitis is most often associated with trauma or is from an endogenous source. Postoperative epidemics from *Candida parapsilosis*[13] and *Paecilomyces lilacinus*[14] were caused by contaminated ir-

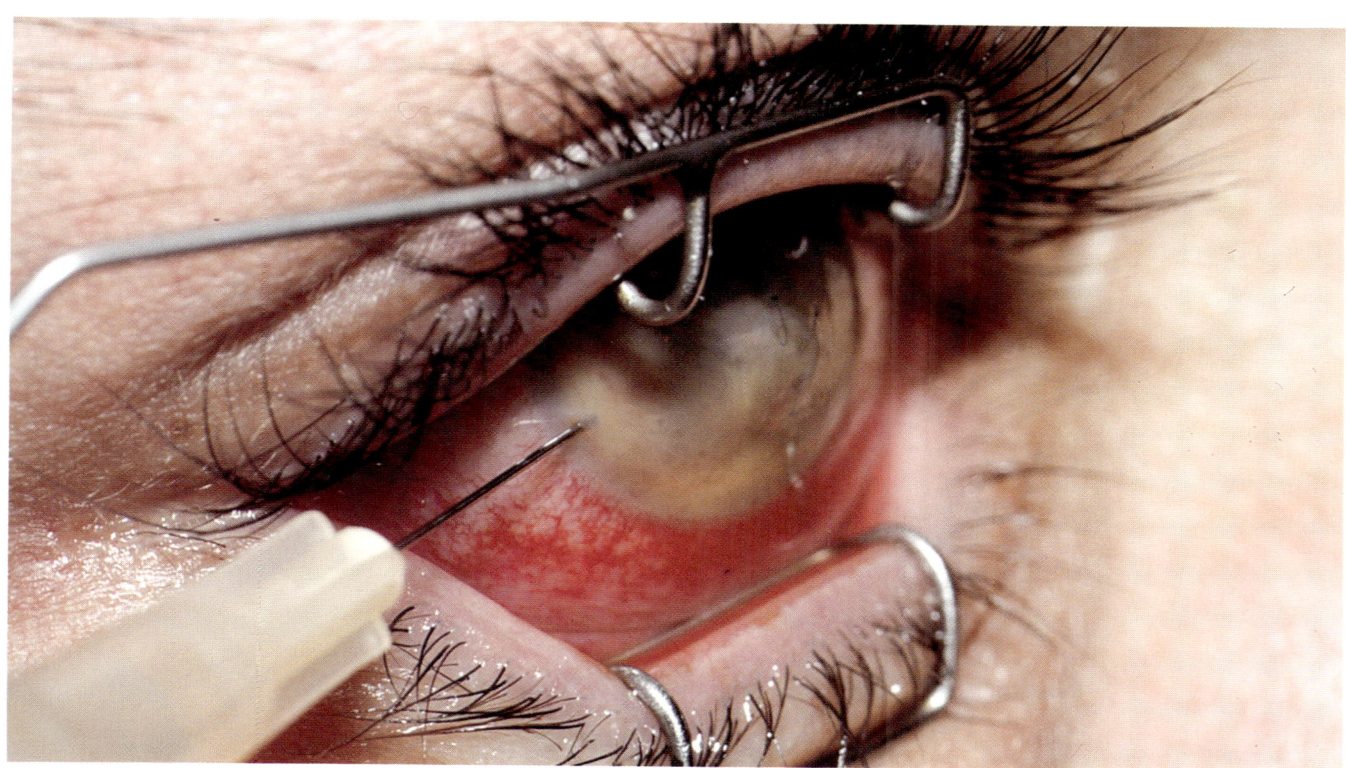

FIGURE 8–1 Anterior chamber paracentesis is performed in a case of endophthalmitis following trauma. The traumatic rupture had been sutured closed a few days earlier.

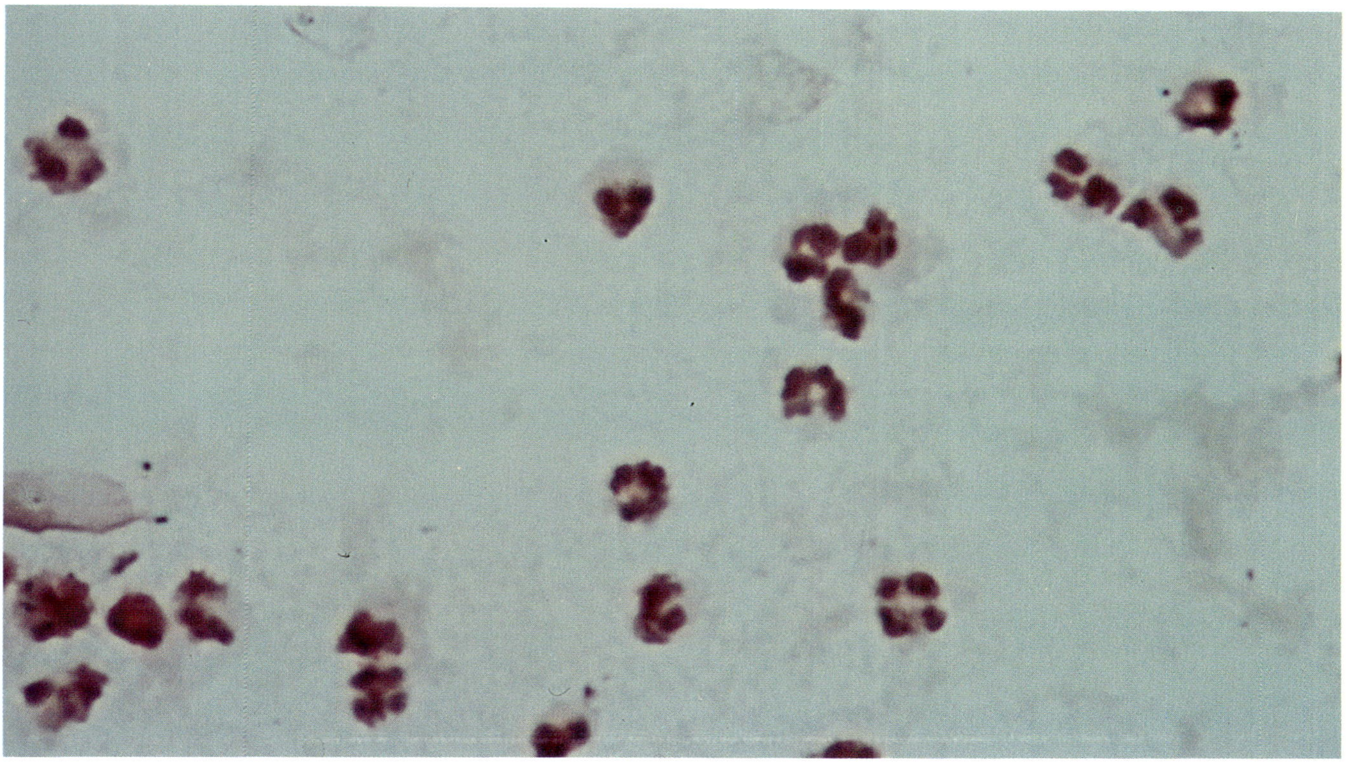

FIGURE 8–2 Aqueous smear from a patient with endophthalmitis shows numerous neutrophils, but no organisms. (Gram stain, × 1000)

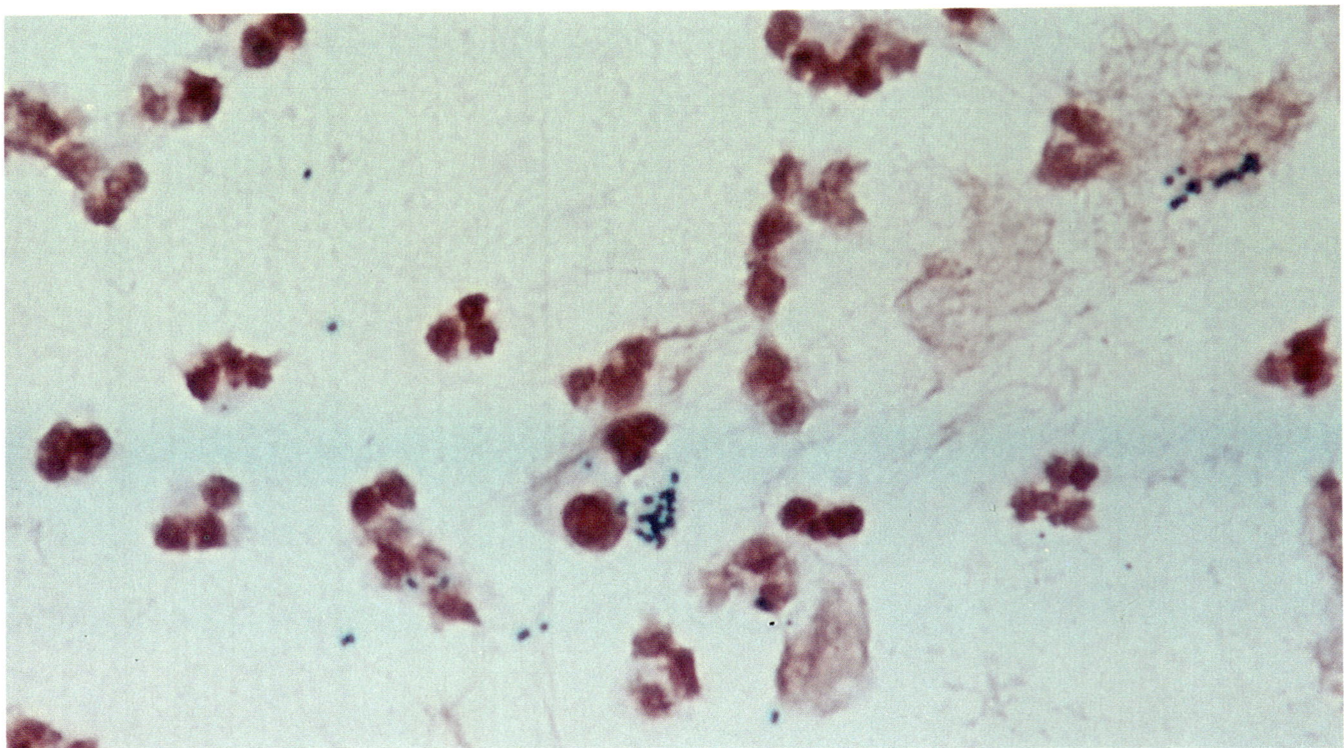

FIGURE 8–3 A vitreous smear from the same patient as shown in Figure 8–2 yields numerous gram-positive cocci. *S. epidermidis* was isolated in culture. (Gram stain, × 1000)

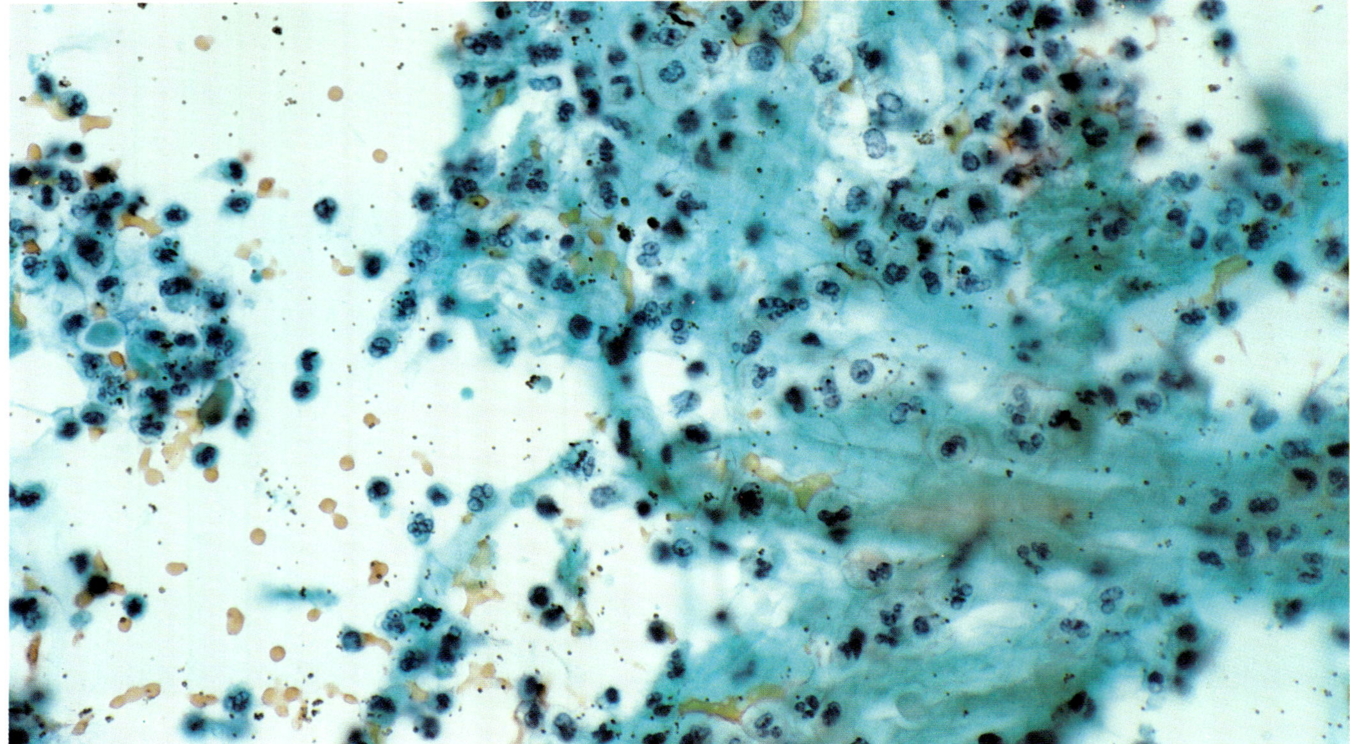

FIGURE 8–4 Cytospin preparations show numerous neutrophils and fibrin. (Papanicolaou, × 250)

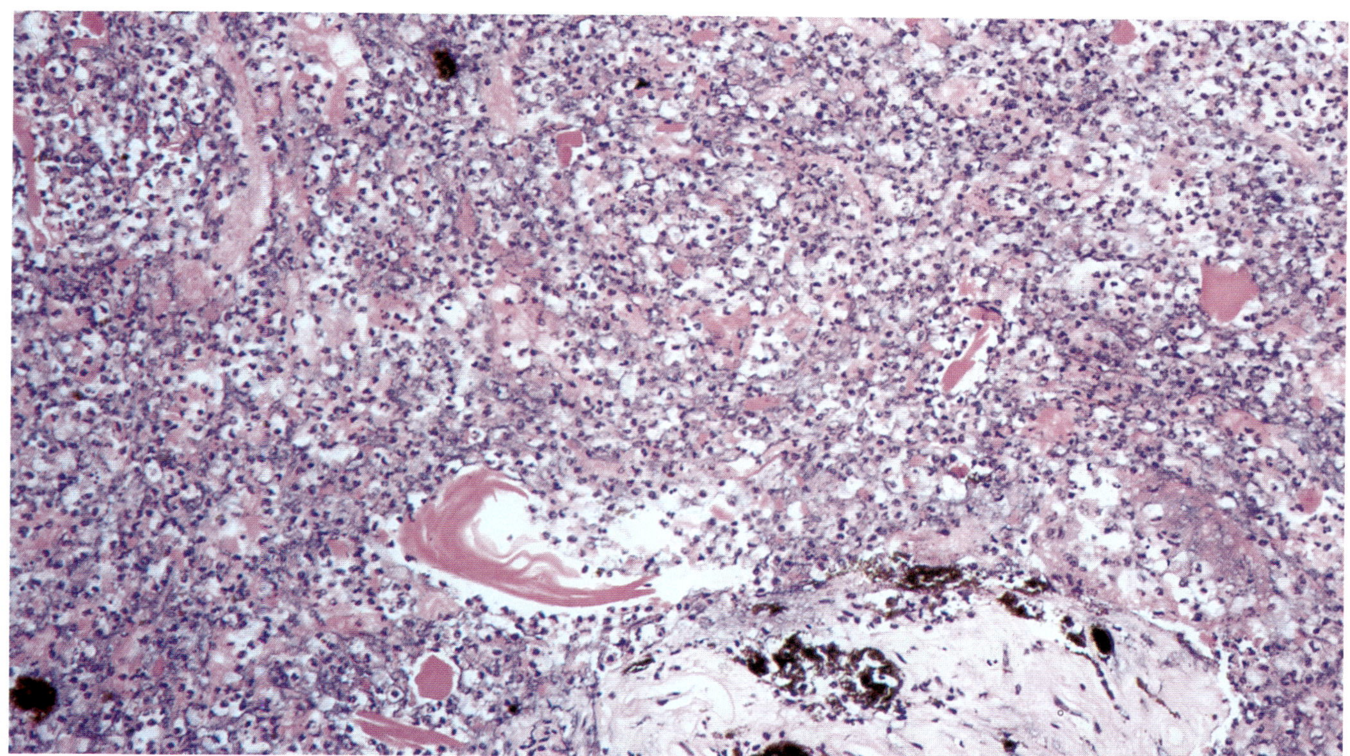

FIGURE 8–5 Cell button shows fragments of normal iris and lens fibers immersed in sheets of neutrophils and necrotic debris. (hematoxylin and eosin, × 125)

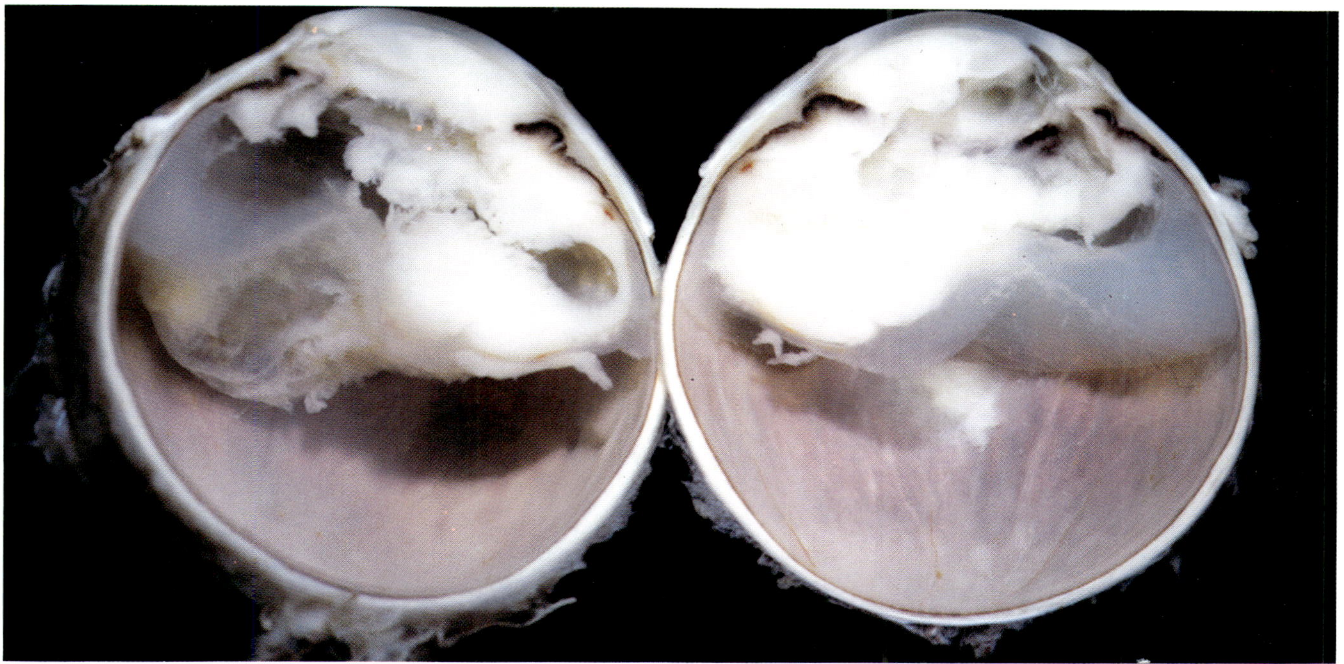

FIGURE 8–6 Gross photograph of bacterial endophthalmitis demonstrates pus filling the anterior chamber and extending into the vitreous cavity. Reprinted with permission of the publisher from Arch Ophthalmol 1980;98:1025–1039. Copyright 1980 American Medical Association.

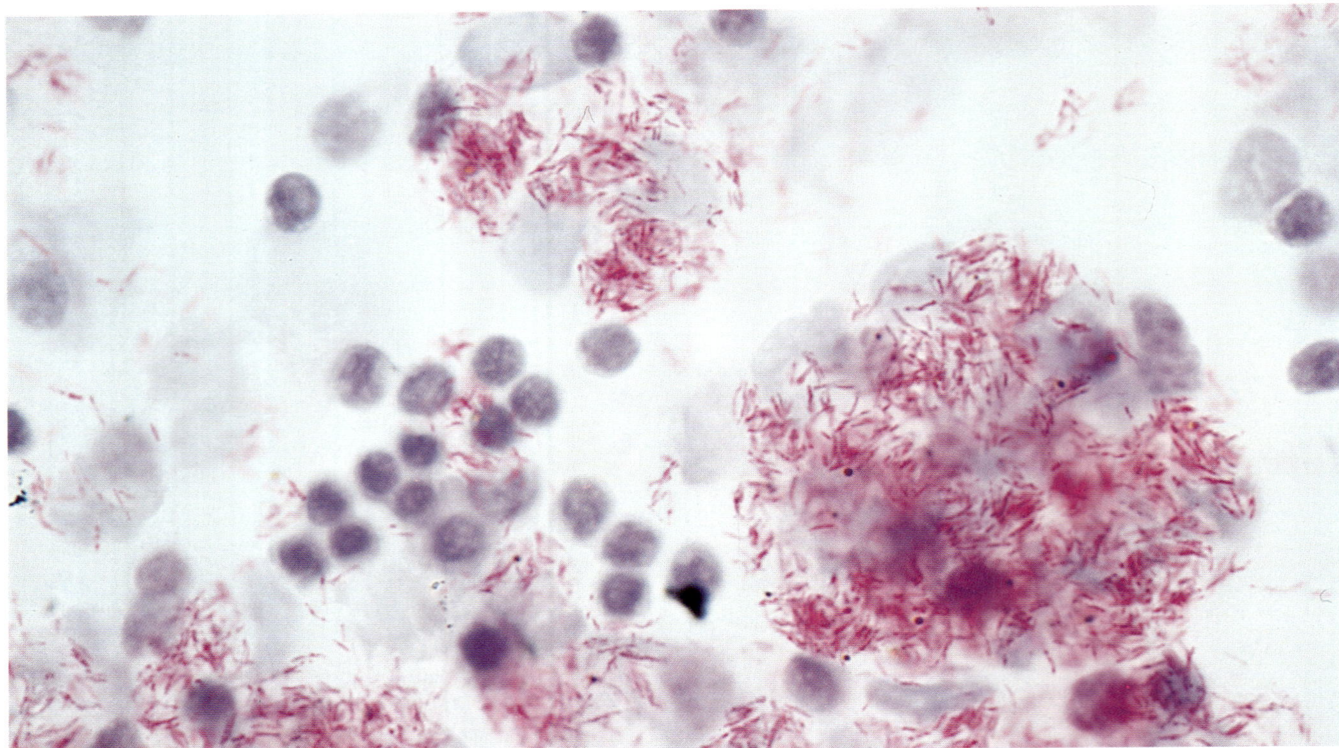

FIGURE 8–7 Cytospin preparation from a 43-year-old female contact-lens wearer who developed an *M. fortuitum* corneal ulcer. Two months after corneal transplant, a dense vitreous opacity and retinal detachment were noted. Intraocular washings showed macrophages stuffed with acid-fast bacilli. (modified Ziel-Nielson, × 720)

rigating solutions. An excellent review of endogenous fungal endophthalmitis has been published.[15]

CANDIDA

Candida endophthalmitis is the most common endogenous fungal infection of the choroid. By the time symptoms are noted, two thirds of patients have bilateral disease and over half have vitreous involvement.[16] About one third of patients with candidemia develop ocular candidiasis.[17,18] Ocular candidemia and intraocular infection is associated with major surgery, bacterial sepsis, systemic antibiotic use, intravenous drug abuse, indwelling catheters, and debilitating illness.[19–22] Vitrectomy can be an excellent means of making the diagnosis.[23] Intraocular washings demonstrate budding yeast with pseudohyphae associated with acute and chronic inflammation (Figure 8–8). The organism is easily identified with periodic acid-Schiff preparations (Figure 8–9).

ASPERGILLUS

Aspergillus endophthalmitis is usually endogenous and is associated with intravenous drug abuse, organ transplantation, and endocarditis.[24–27] The infection spreads presumably by hematogenous seeding with deposition in the choroid and retina and extends secondarily into the vitreous cavity. Vitrectomy specimens reveal septated hyphae, which branch at 45° angles. Gomori methenamine silver stains will highlight the fungal elements (Figure 8–10).

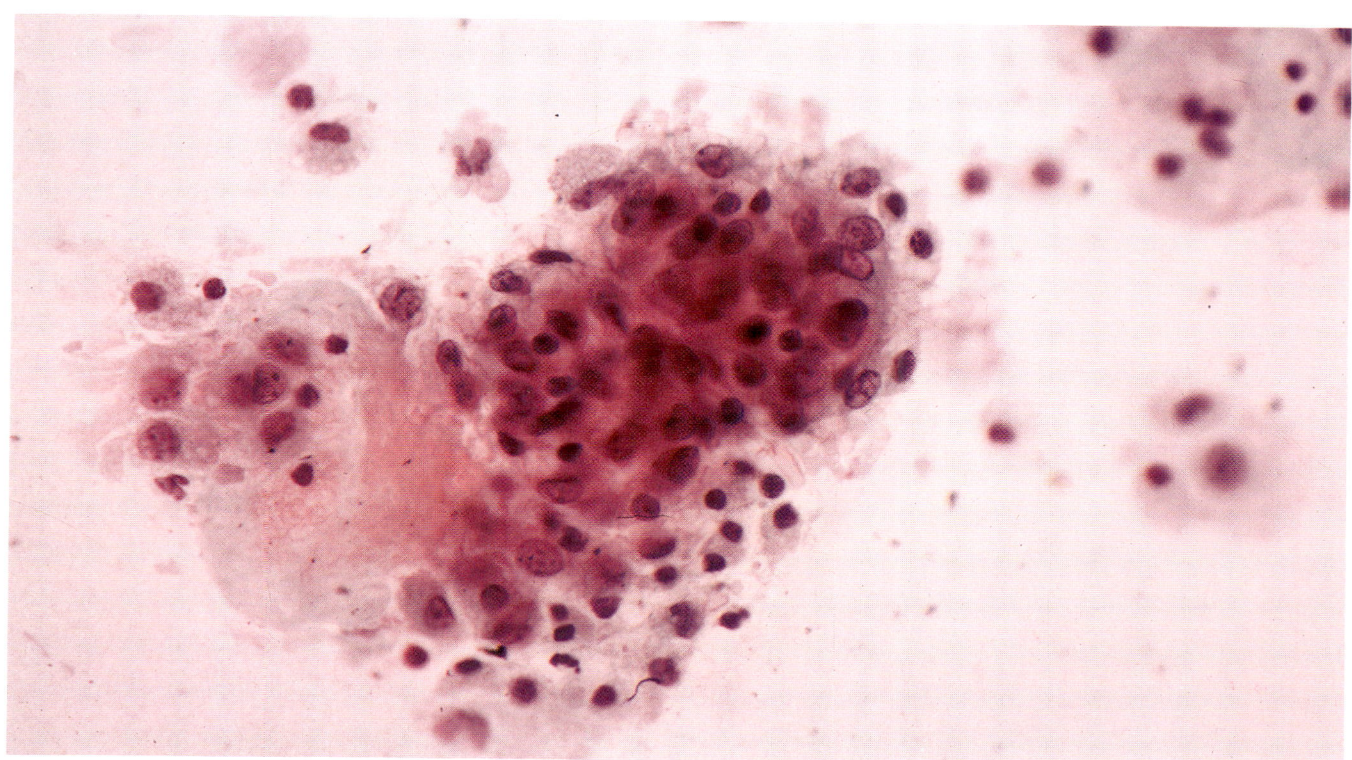

FIGURE 8–8 Cytospin preparation from intraocular washing demonstrates granulomatous inflammation from a case of culture-proven *Candida* endophthalmitis.

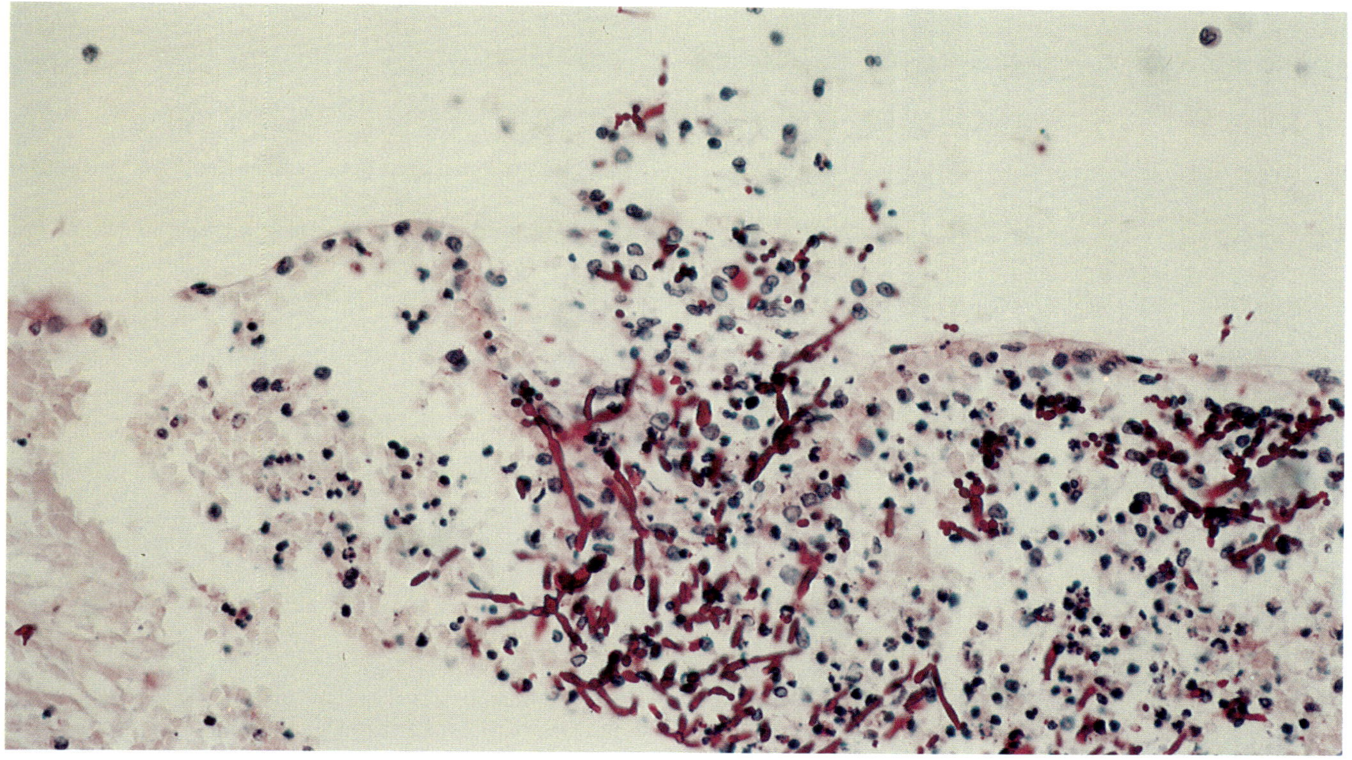

FIGURE 8–9 Histologic section at the vitreoretinal interface shows periodic acid-Schiff positive budding yeast and pseudohyphae.

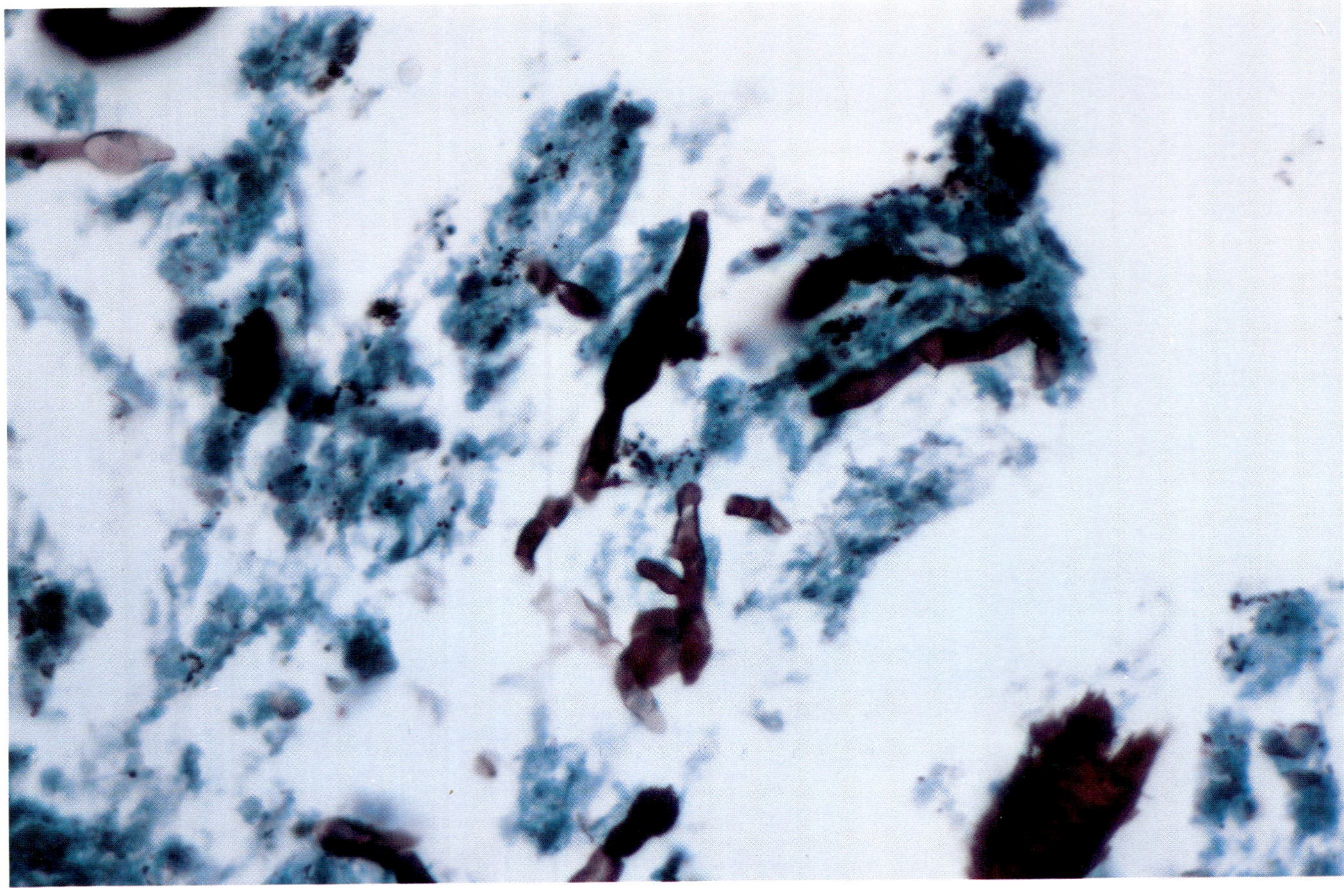

FIGURE 8–10 Intraocular washing from a 38-year-old intravenous drug user shows septated branching hyphae. *A. flavus* was isolated in culture. (Gomori methenamine silver, × 720)

FUSARIUM

Fusarium solani is one of the most common organisms identified in fungal keratitis.[28,29] It is one of the nonpigmented filamentous fungi. It has a propensity to invade the cornea, penetrate Descemet's membrane, and involve the anterior and posterior chambers.[30] In the course of diagnosis, cytologic specimens, including corneal scraping, anterior chamber aspirates, and intraocular washings, may be obtained. Silver staining reveals septated hyphal forms that are difficult to differentiate morphologically from *Aspergillus* (Figure 8–11).

COCCIDIOIDOMYCOSIS

Coccidioides immitis is a dimorphic fungus that is endemic in the soil of the San Joaquin Valley (California), Arizona, New Mexico, and Texas. Intraocular coccidioidomycosis usually results as either anterior uveitis (iritis and iridocyclitis) or posterior uveitis.[31–33] Infections of the choroid and retina appear to be a result of hematogenous seeding.[34] In some cases of anterior uveal involvement, the diagnosis may be confirmed by anterior chamber paracentesis.[35,36] Smears show granulomatous inflammation and spherules containing endospores that are visible in hematoxylin and eosin stained material, but are greatly enhanced by silver preparations (Figure 8–12).

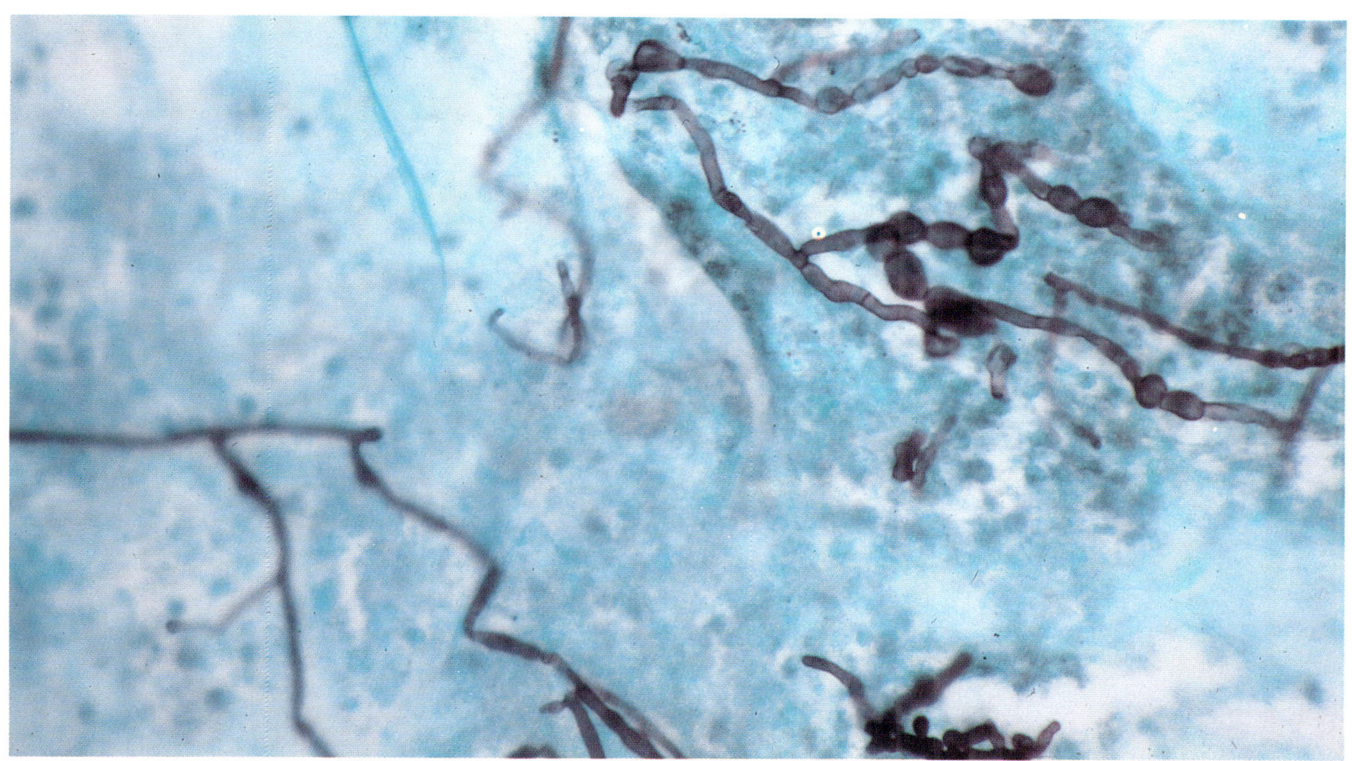

FIGURE 8–11 Intraocular washing from a patient with keratitis and endophthalmitis. Hyphal forms with septae are visible. *F. solani* grew in culture. (Gomori methenamine silver, × 480)

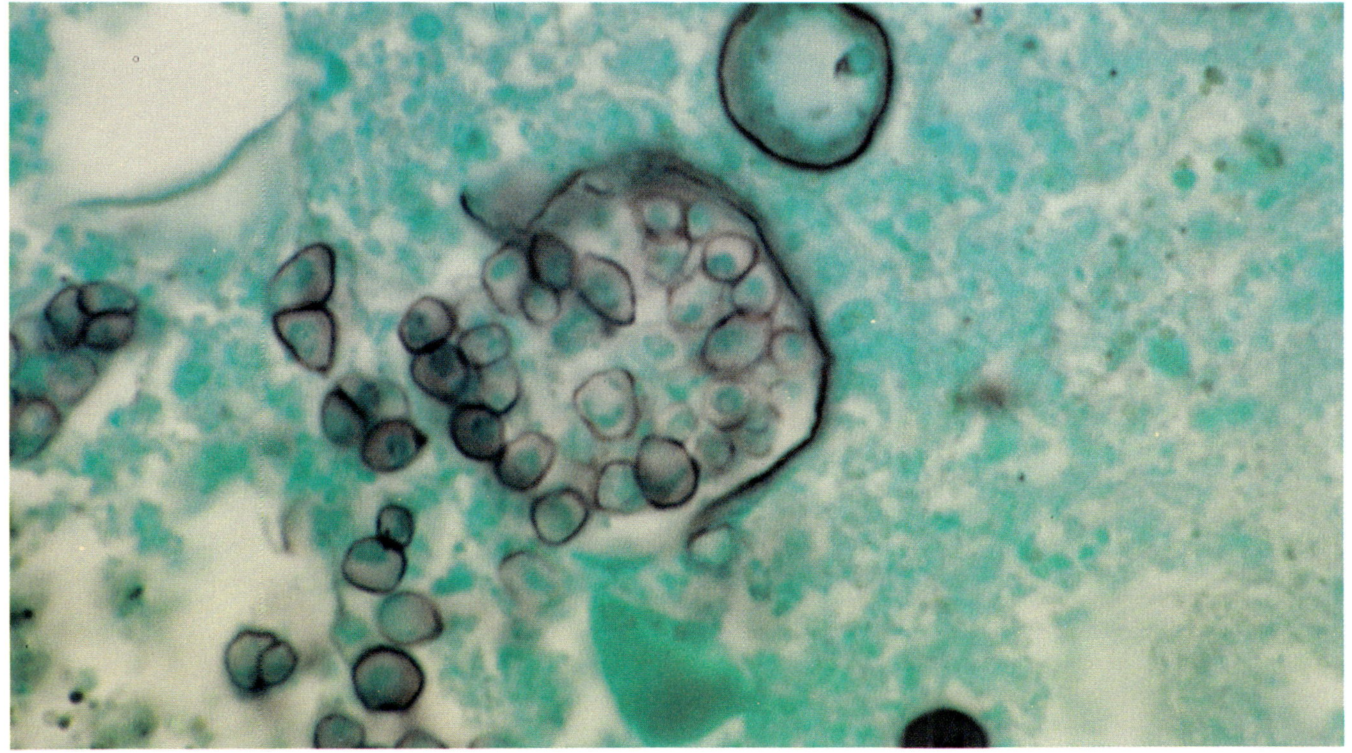

FIGURE 8–12 Cytospin preparation of intraocular washing demonstrates spherules and endospores in *C. immitis* infection. (periodic acid-Schiff, × 720)

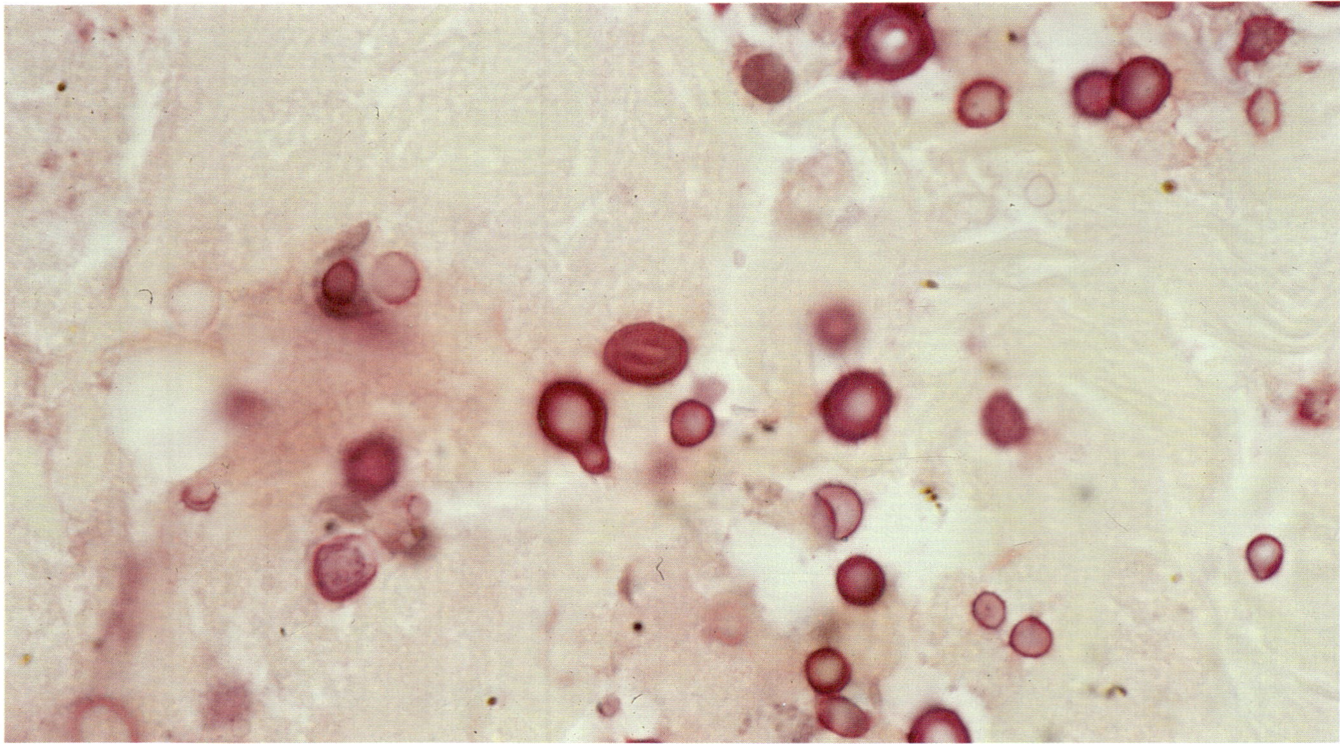

FIGURE 8–13 Cytospin preparation shows small budding yeast indicative of *Cryptococcus*. (mucicarmine, × 720)

CRYPTOCOCCUS

Ocular *cryptococcus* may be associated with meningitis in otherwise healthy individuals or may be a manifestation of a disseminated infection in immunocompromised hosts.[37, 38] It presents as a multifocal chorioretinitis with vitreous involvement.[39, 40] The vitreous may be involved by direct extension from the uvea. Intraocular washings show budding yeast with a thick mucicarmine and periodic acid-Schiff staining capsule (Figure 8–13). The yeast forms vary in size. Associated inflammation is usually mild in immunocompromised hosts or may be intensely suppurative and granulomatous in immunocompetent individuals.

CYSTICERCUS CELLULOSAE (TAENIA SOLIUM)

Cysticercosis is the most common ocular tapeworm infection. *Cysticercus cellulosae* is the larval stage and *Taenium solium* is the adult worm. *T. solium* is endemic in Eastern Europe, Africa, Asia, and Central and South America. Infection occurs in man when raw pork is ingested and the larvae grow into the adult worm. Ocular cysticercosis occurs if eggs are regurgitated or ingested and develop into larvae that penetrate the small intestine. The larvae may lodge in the many ocular structures including the anterior chamber, vitreous cavity, subretinal space, and subconjunctival space.[41–47] Pars plana vitrectomy is the currently recommended procedure for removing the cysts from the vitreous cavity.[48–51]

The intact worm is easily recognized with its scolex, rostellum, suckers, and hooklets.[52] However, because of sampling bias, specimens removed with fine needle aspiration or vitrectomy may include only fragments of the cyst wall or body of the worm and the diagnosis is thereby much more difficult (Figures 8–14 and 8–15).[53]

CYTOMEGALOVIRUS

Cytomegalovirus (CMV) produces a characteristic necrotizing retinitis in the eyes of immunocompromised adults and infants (Figure 8–16).[54] CMV retinitis is the most common significant ocular infection in the acquired immunodeficiency syndrome (AIDS).[55] Initially, it manifests as a perivascular distribution, producing a distinctive granular appearance (Figures 8–17 and 8–18).[56] Necrotizing infection may lead to retinal detachment.[57,58] Intraocular washings from attempted surgical repair may contain fragments of infected retina. Cytologic confirmation of CMV infection is frequently quite helpful in management. Characteristic intranuclear and intracytoplasmic inclusions by light microscopy and dense bodies and viral particles by electron microscopy are identified (Figure 8–19).

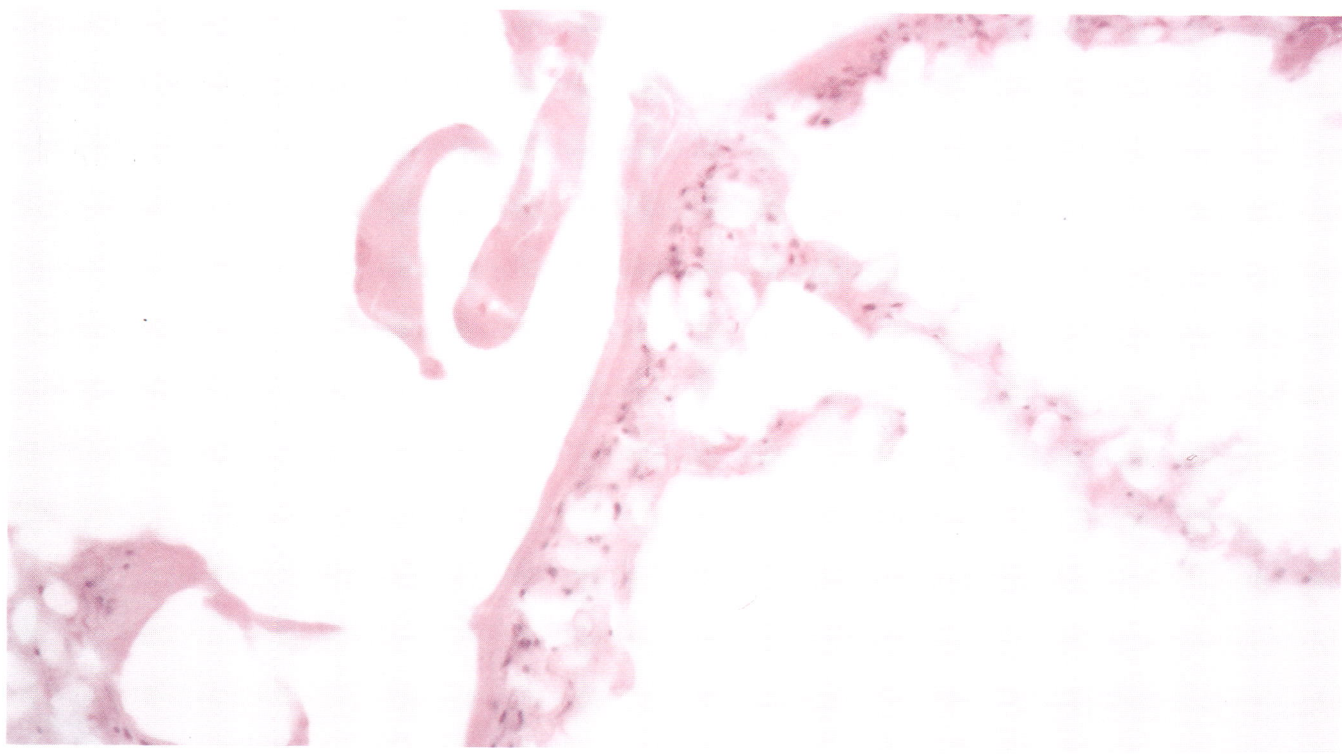

FIGURE 8–14 (ABOVE) Photomicrograph of the cytospin preparation shows a cyst wall of *T. solium* as a thin acellular cuticular membrane with an eosinophilic cuticle. (hematoxylin and eosin, × 250)

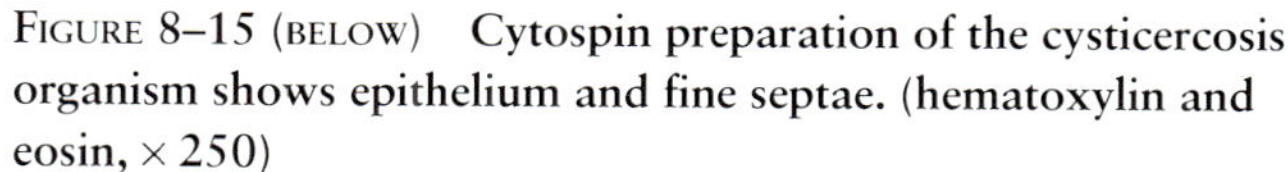

FIGURE 8–15 (BELOW) Cytospin preparation of the cysticercosis organism shows epithelium and fine septae. (hematoxylin and eosin, × 250)

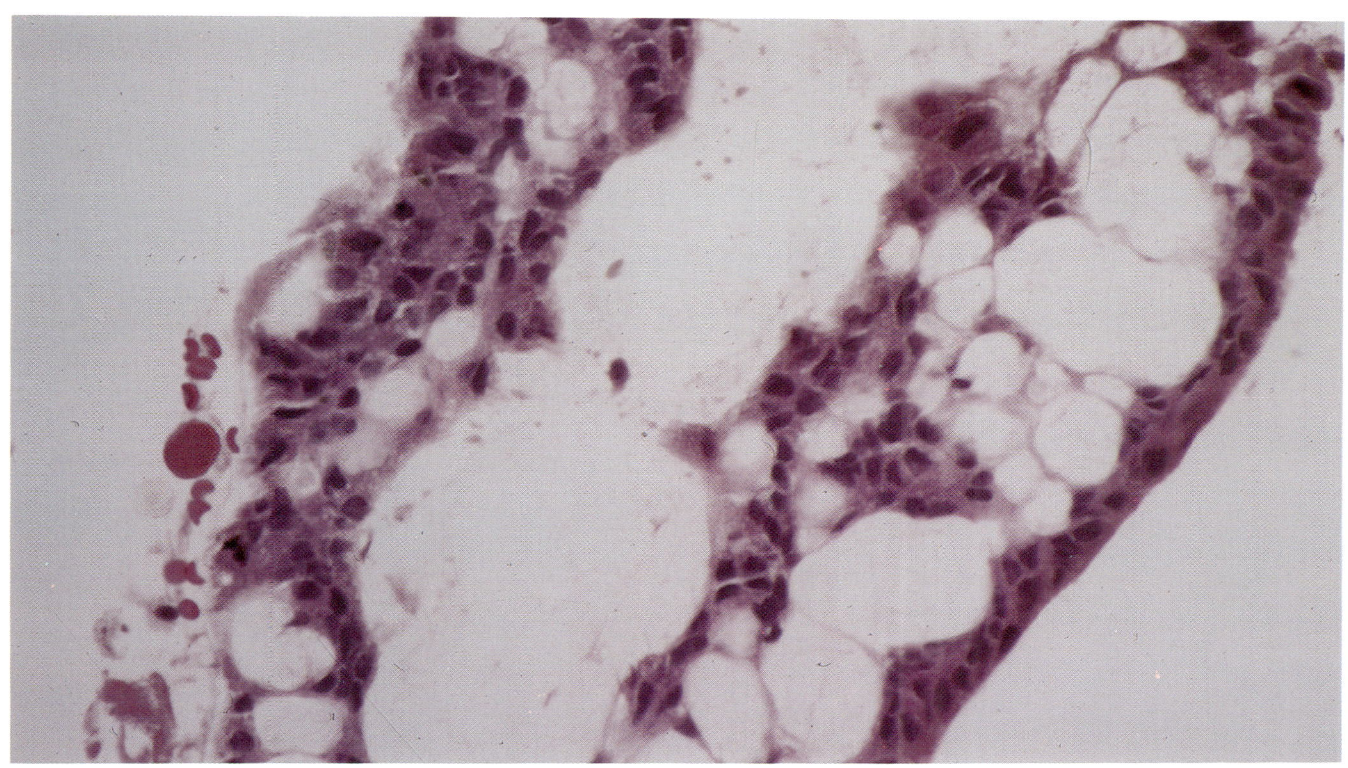

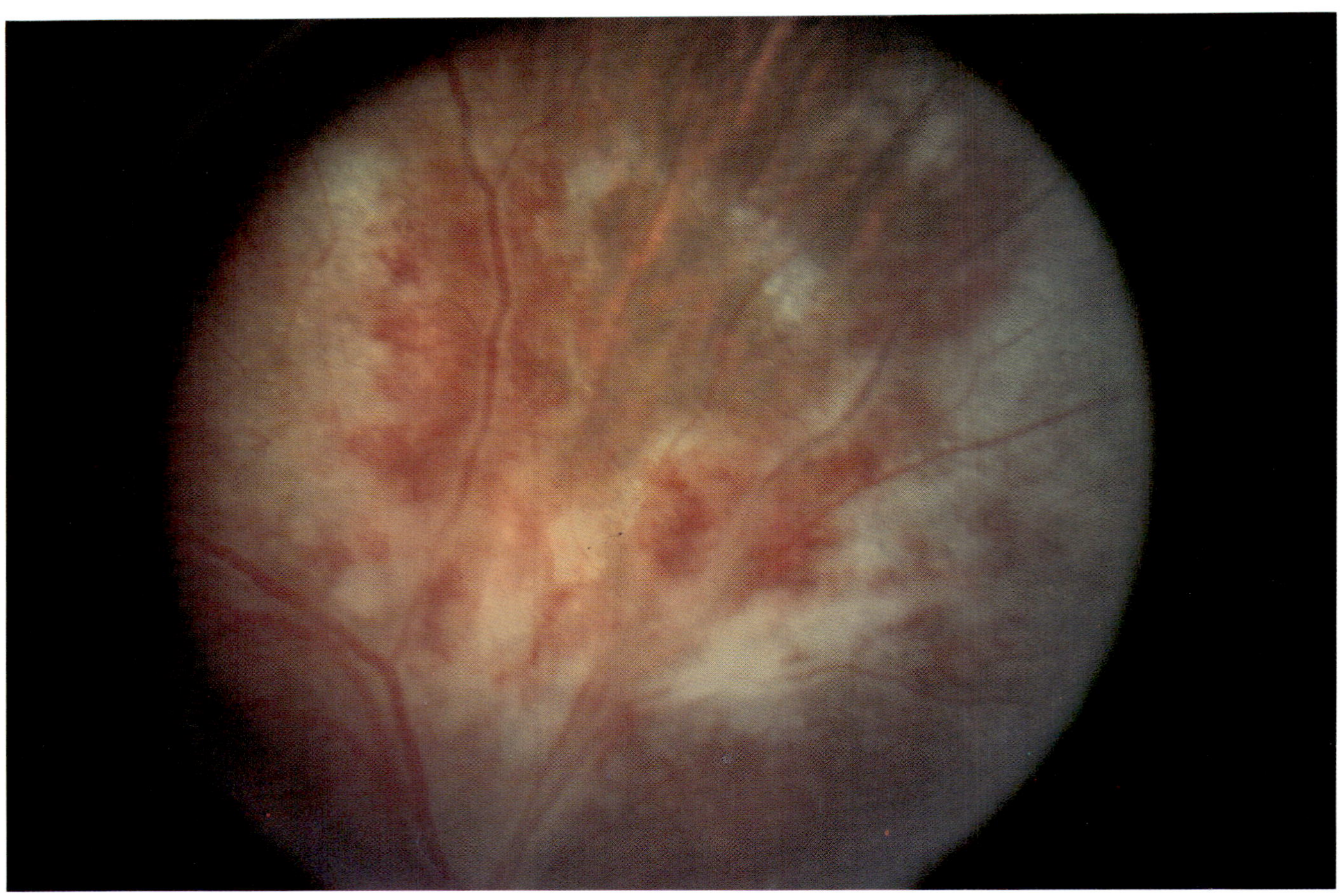

FIGURE 8–16 Clinical photograph demonstrates hemorrhage and retinal necrosis in severe CMV retinitis in a patient with AIDS.

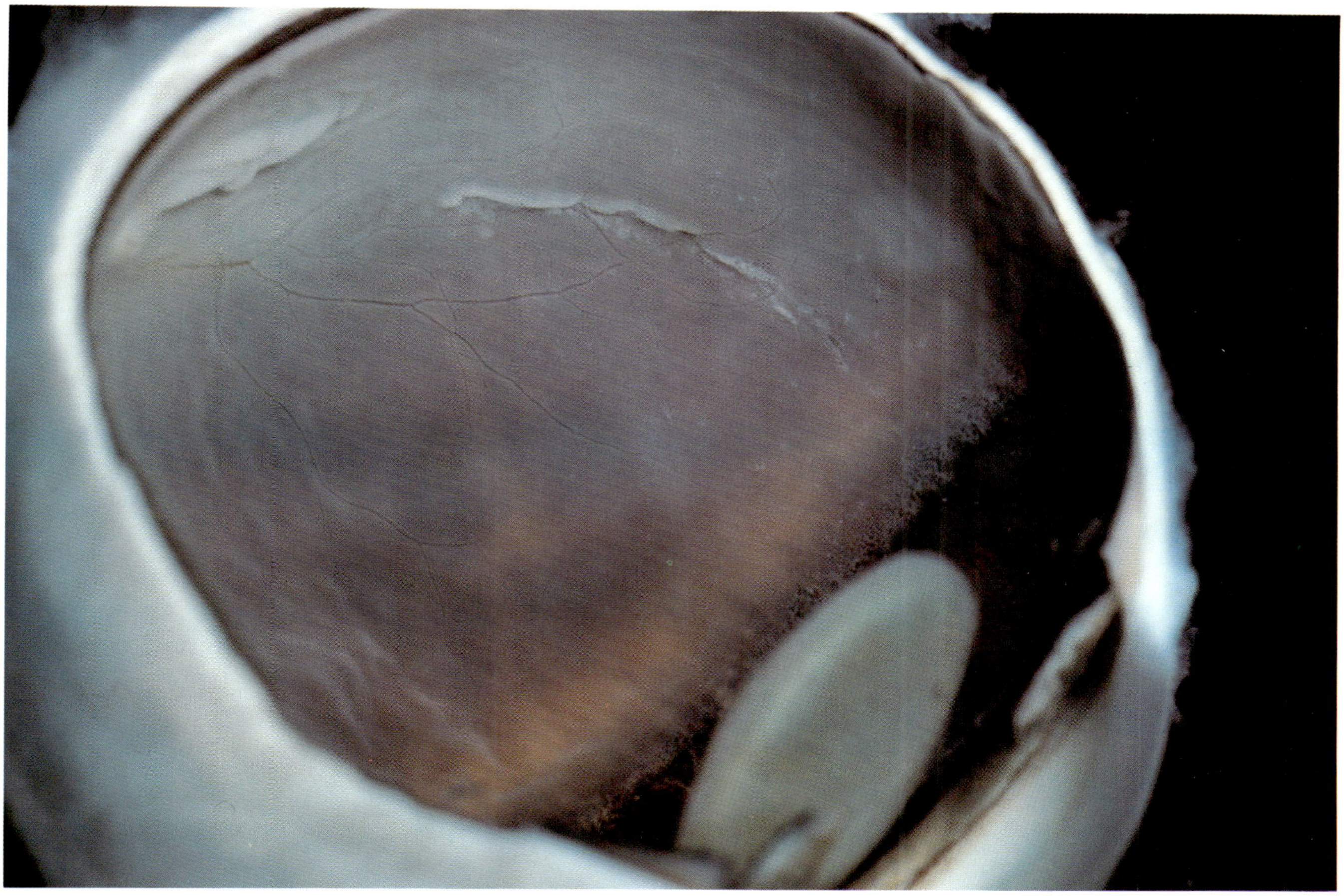

FIGURE 8–17 (ABOVE) Gross photograph illustrates retinal granular lesions from early CMV retinitis.

FIGURE 8–18 (BELOW) A closer view of Figure 8–17 shows that the white granular necrosis follows a perivascular distribution.

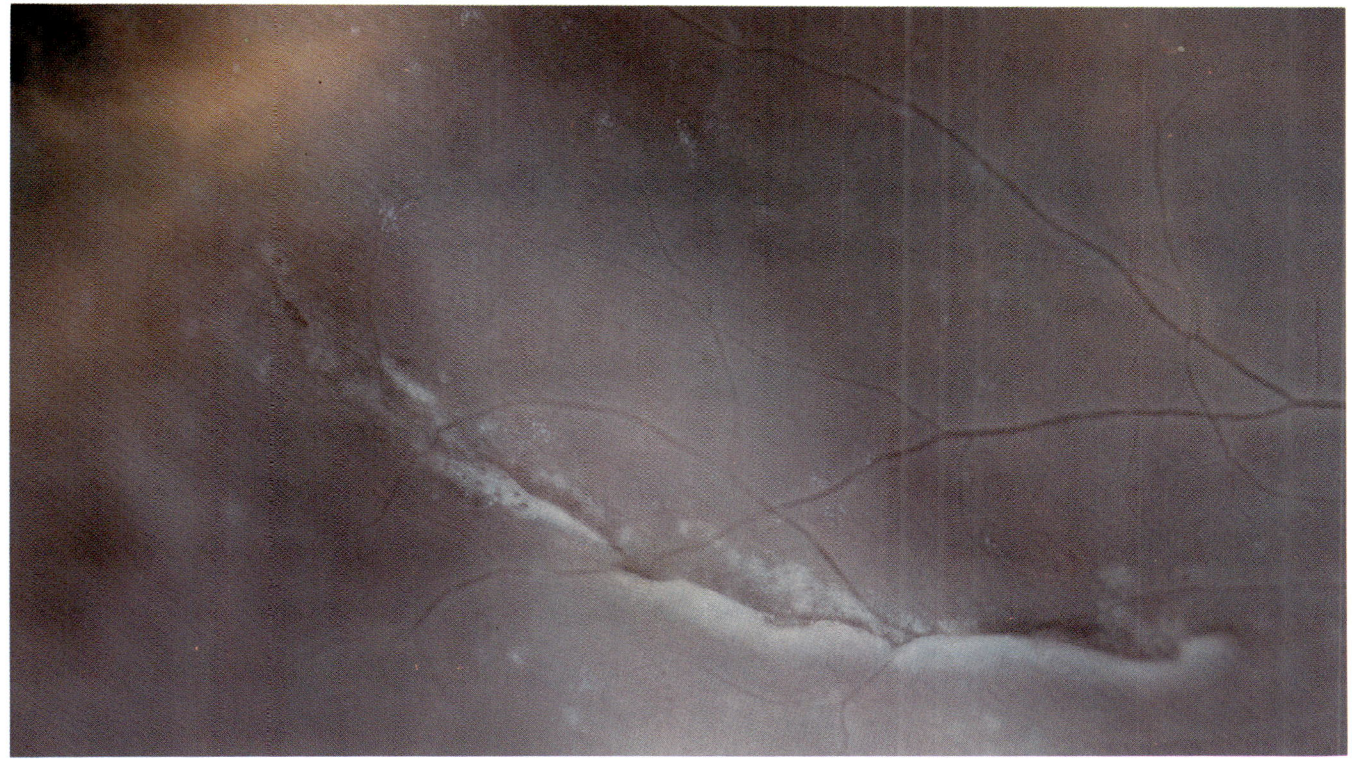

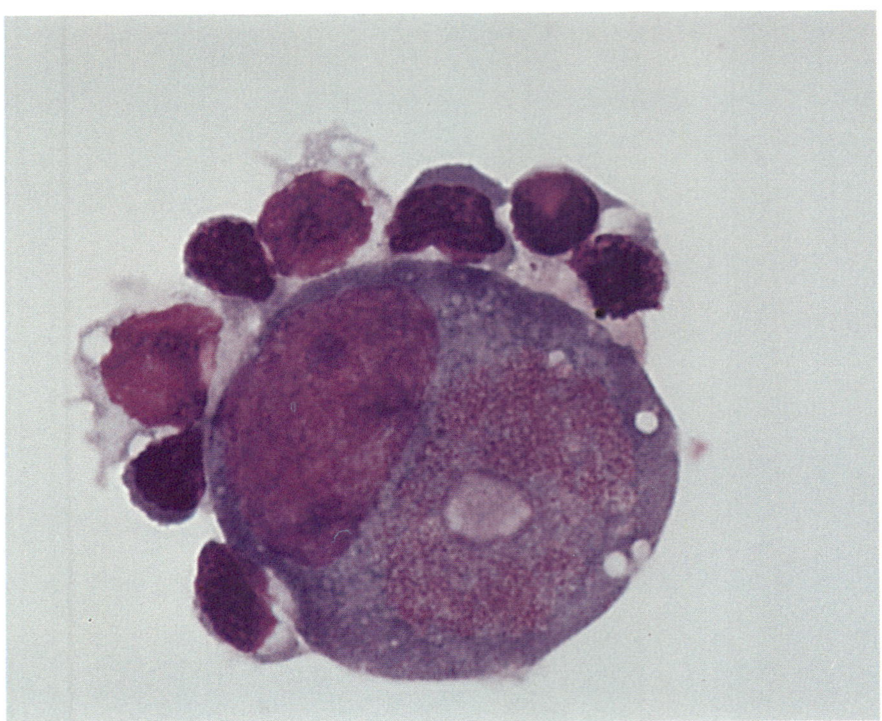

FIGURE 8–19 Cytospin preparations show necrotic retina with inclusions within the enlarged cells. (May-Grünwald Giemsa, × 720)

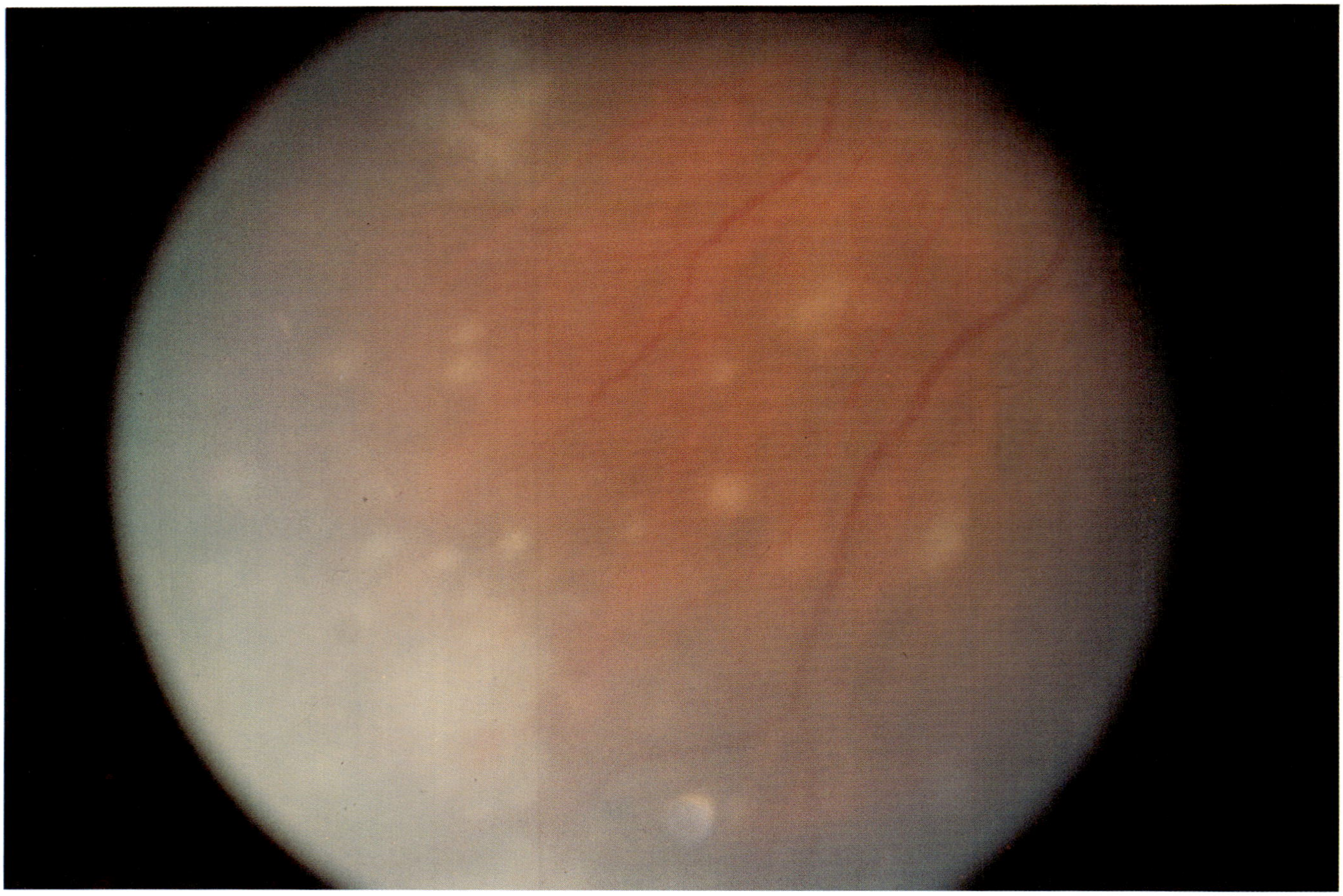

FIGURE 8–20 Fundus photograph of a previously health 37-year-old male shows vitreous haze and multifocal yellow-white necrotic foci in acute retinal necrosis syndrome.

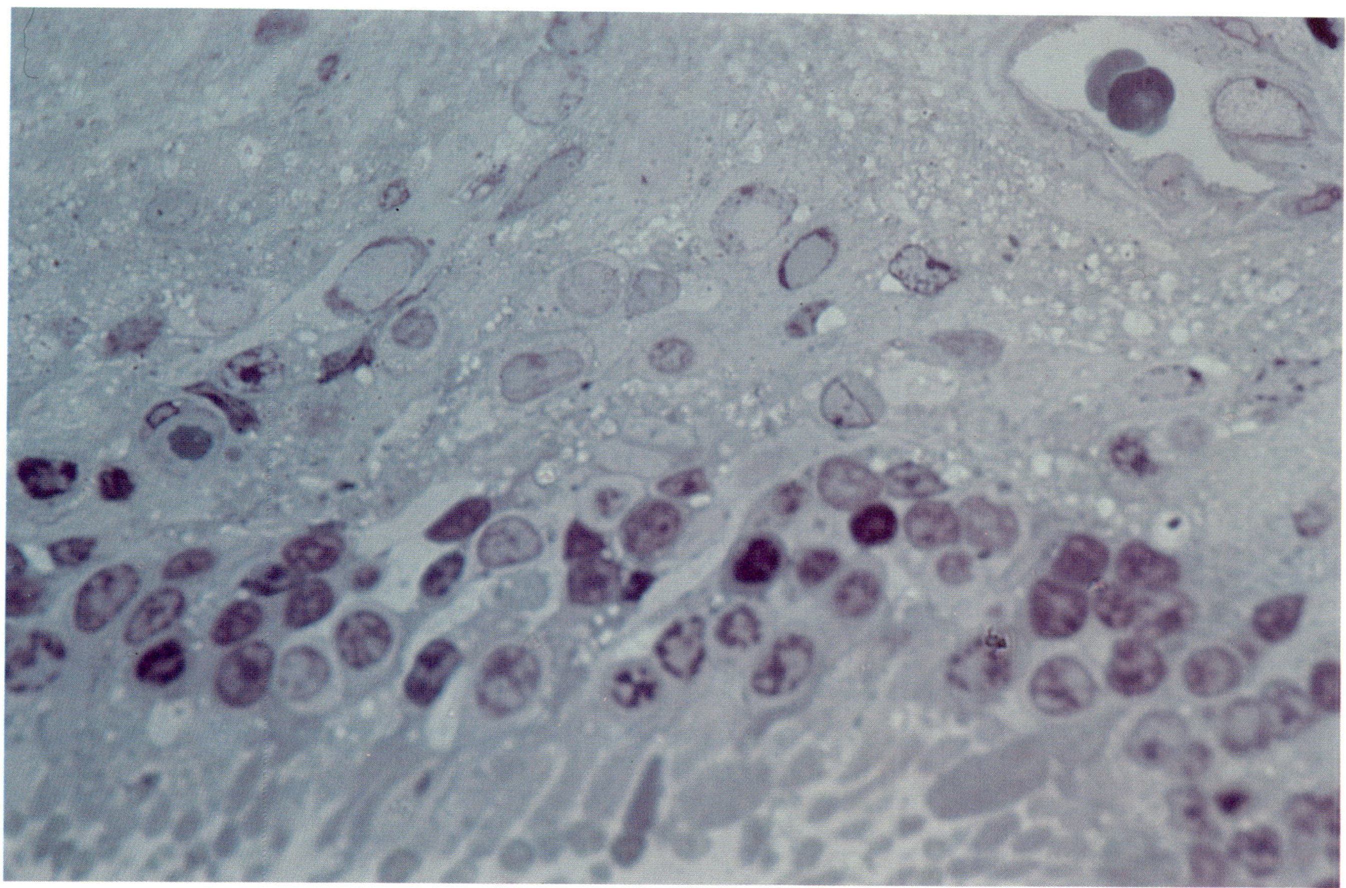

FIGURE 8–21 Retinal fragment, processed for electron microscopy, from a patient with a retinal detachment, and the acute retinal necrosis syndrome shows necrotic retinal fragment with ground glass intranuclear inclusions. (toluidine blue one micron section of cell button, × 125)

ACUTE RETINAL NECROSIS SYNDROME

In otherwise healthy adults, a syndrome of acute necrotizing retinitis, vitritis, and vasculitis has been associated with varicella zoster virus[59–61] and herpes simplex virus (Figure 8–20).[62–66] Histologic studies illustrate full thickness retinal necrosis, underlying chronic choroiditis, and dense perivasculitis.[67] Retinal detachment may be caused by vitreous traction alone or combined with retinal breaks (rhegmatogenous retinal detachment).[68] Intraocular washings show severe necrosis and acute inflammation. Fragments of retina are infiltrated by neutrophils with only ghost outlines of their neural fibers. Ground glass intranuclear inclusions may be visible (Figure 8–21).

TOXOPLASMOSIS

Toxoplasma gondii may occur in congenital or acquired forms. Typically, the infants become congenitally infected only if the mother acquires the disease immediately prior to or during gestation. If the infection occurs late in pregnancy, the congenital manifestations may be asymptomatic. In these cases, the disease may only become apparent as a chorioretinal scar in childhood (Figure 8–22). Acquired toxoplasmosis is usually mild in immunocompetent hosts. However, in immunocompromised hosts, the infection may produce a severe necrotizing retinitis.[69–72] If the diagnosis is difficult, a vitrectomy and retinal biopsy may be performed. Intraocular washings may reveal cysts with typical bradyzoites (Figure 8–23). Histopathology confirms the presence of a severe necrotizing infection with organisms found in the retina (Figure 8–24).[73]

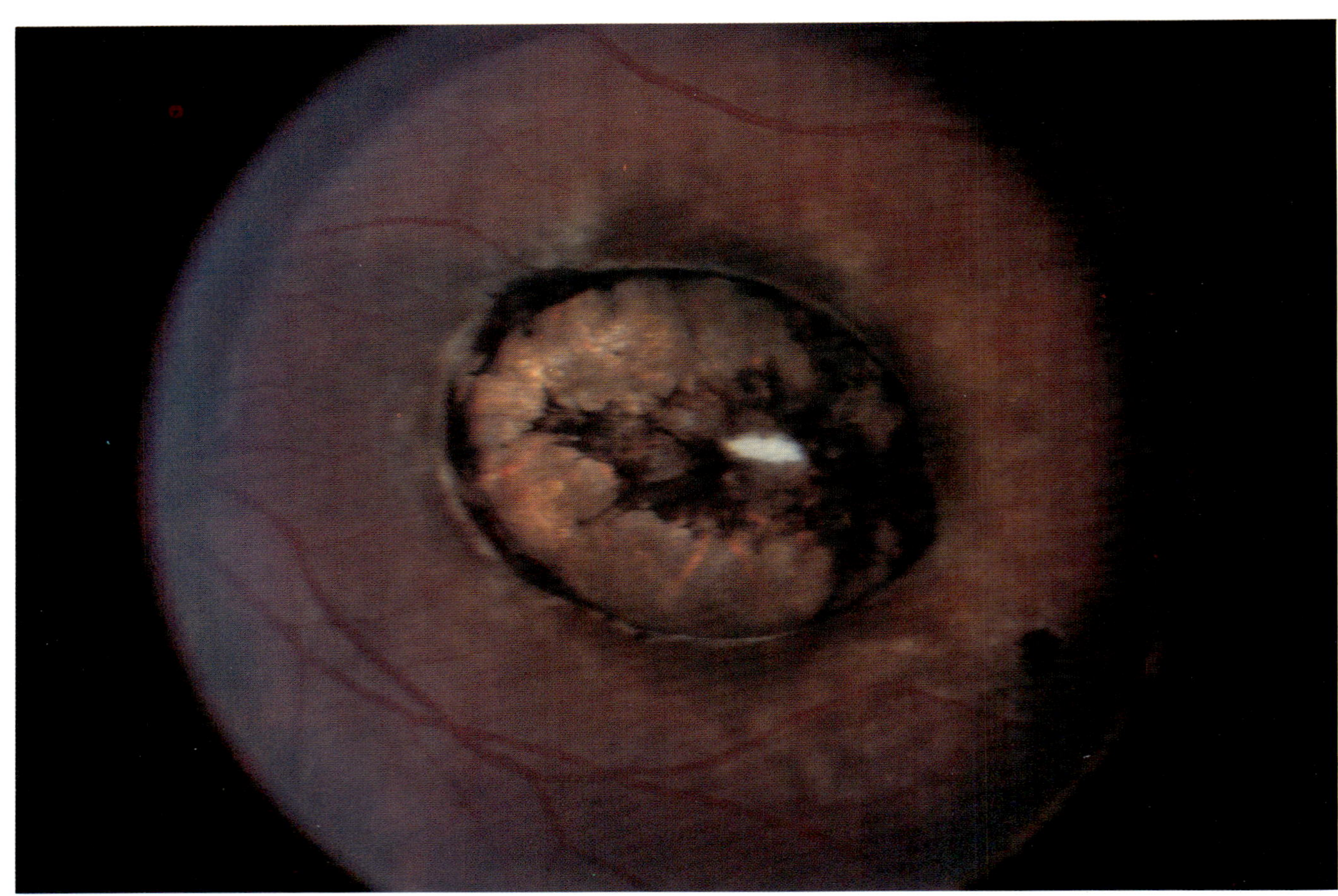

FIGURE 8–22 Clinical photograph shows dense macular chorioretinal scar from a patient who had congenital toxoplasmosis.

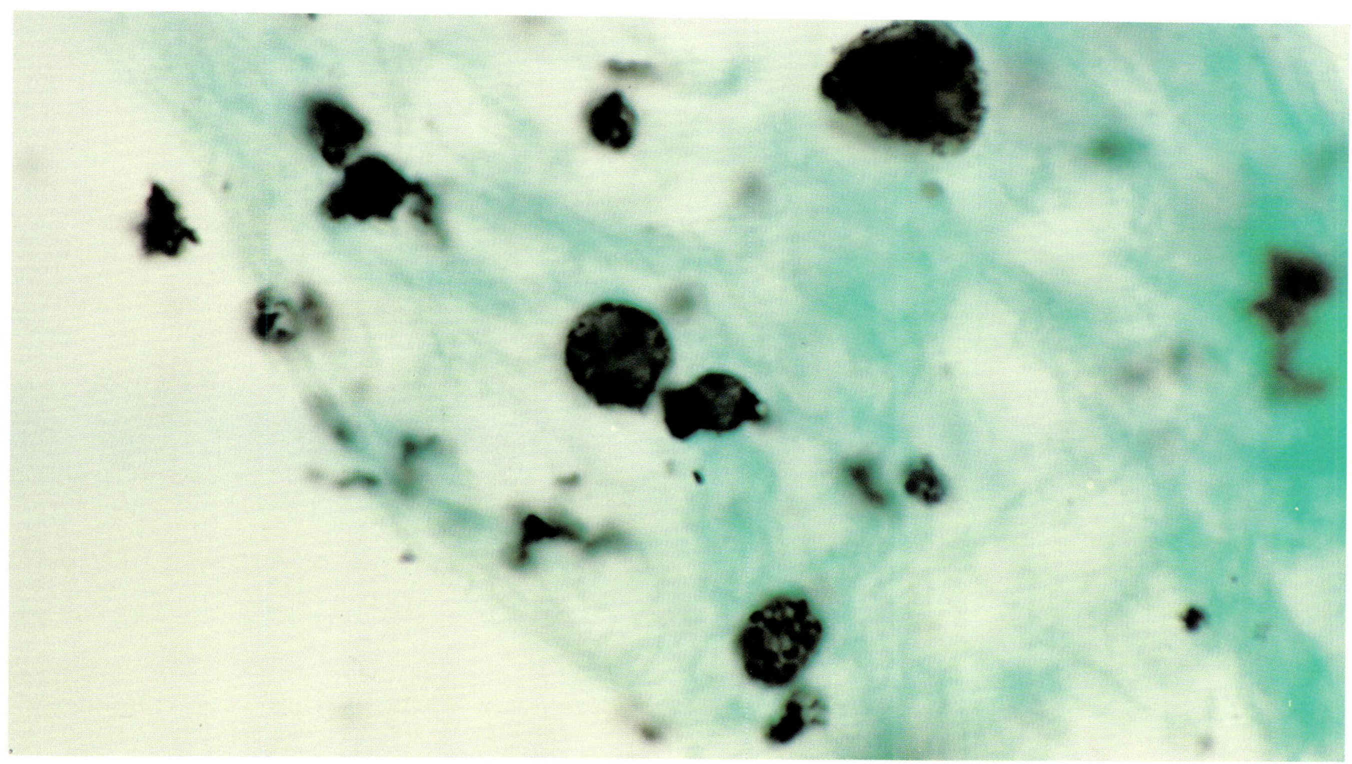

FIGURE 8–23 Cytospin preparation shows encysted bradyzoites. (Gomori methenamine silver, × 720)

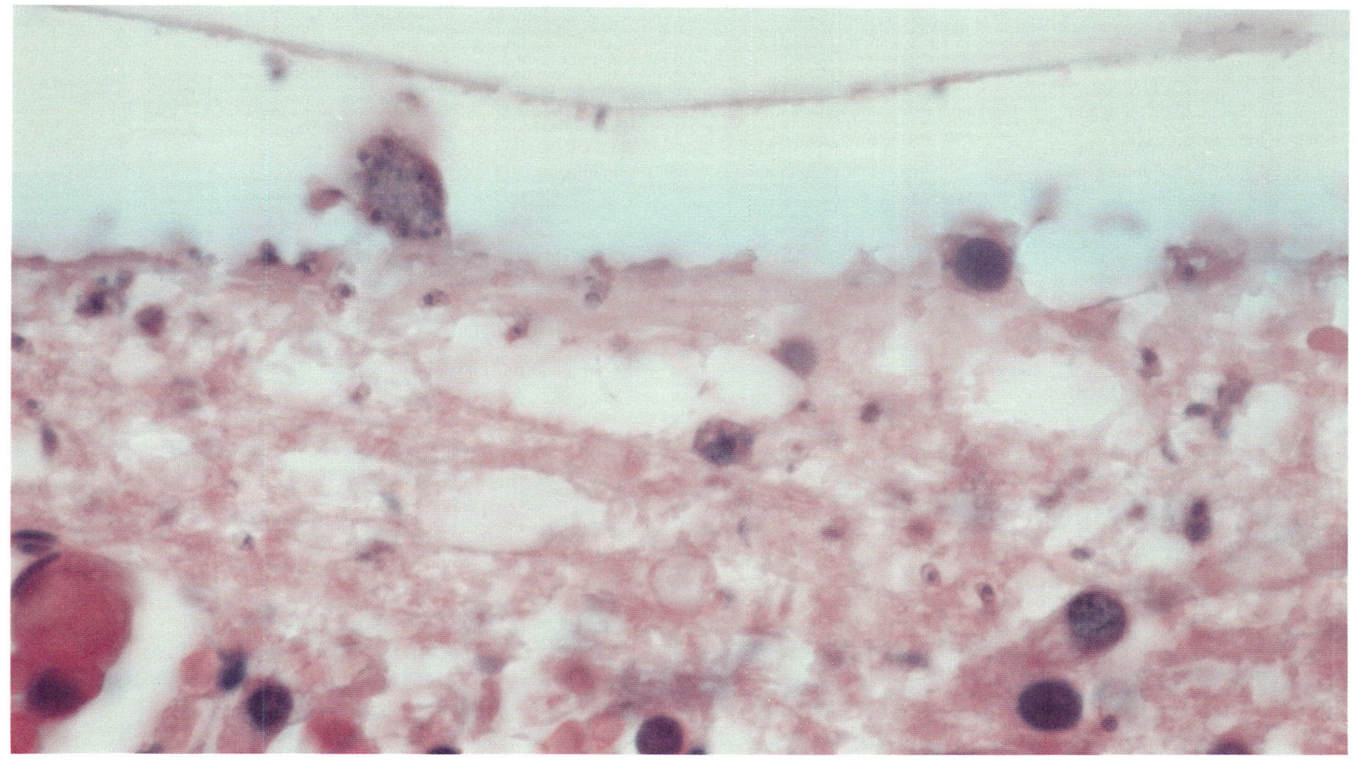

FIGURE 8–24 Histologic section shows cysts and bradyzoites of *T. gondii* in the retina. (hematoxylin and eosin, × 500)

REFERENCES

1. Meredith TA. Clinical microbiology of infectious endophthalmitis. In: Ryan SJ, Ogden TE, Schachat AP, eds. Retina. St Louis: C.V. Mosby, 1989;1:183–188.

2. Ficker LA, Meredith TA, Wilson LA, Kaplan HJ, Kozarsky AM. Chronic bacterial endophthalmitis. Am J Ophthalmol 1987;103:745–748.

3. Driebe WT Jr, Mandelbaum S, Forster RK, Schwartz LK, Culbertson WW. Pseudophakic endophthalmitis: diagnosis and management. Ophthalmology 1986;93:442–448.

4. Beatty RF, Fobin JB, Trousdale MD, Smith RE. Anaerobic endophthalmitis caused by *Propionibacterium acnes*. Am J Ophthalmol 1986;101:114–116.

5. Mandelbaum S, Forster RK, Gelender H, Culbertson W. Late onset endophthalmitis associated with filtering blebs. Ophthalmology 1985;92:964–972.

6. Greenwald MJ, Wohl LG, Sell CH. Metastatic bacterial endophthalmitis: a contemporary reappraisal. Surv Ophthamol 1986;31:81–101.

7. Ficker LA, Meredith TA, Wilson LA, Kaplan HJ. Role of vitrectomy in *Staphylococcus epidermidis* endophthalmitis. Br J Ophthalmol 1988;72:386–389.

8. Turner L, Stinson I. *Mycobacterium fortuitum* as a cause of a corneal ulcer. Am J Ophthalmol 1965;60:329–331.

9. Levenson DS, Harrison CH. *Mycobacterium fortuitum* corneal ulcer. Arch Ophthalmol 1966;75:189–191.

10. Zimmerman LE, Turner L, McTigue JW. *Mycobacterium fortuitum* infection of the cornea. Arch Ophthalmol 1969;82:596–601.

11. Wunsh SE, Boyle GL, Leopold IH, Littman ML. *Mycobacterium fortuitum* infection of corneal graft. Arch Ophthalmol 1969;82:602–607.

12. Lazar M, Nemet P, Bracha R, Campus A. *Myco-bacterium fortuitum* keratitis. Am J Ophthalmol 1974;78:530–532.

13. Stern WH, Tamura E, Jacobs RA, Pons VG, Sone RD, et al. Epidemic postsurgical *Candida parapsilosis* endophthalmitis: clinical findings and management of 15 consecutive cases. Ophthalmology 1985;92:1701–1709.

14. Pettit TH, Olson RJ, Foos RY, Martin WJ. Fungal endophthalmitis following intraocular lens implantation: a surgical epidemic. Arch Ophthalmol 1980:98:1025–1039.

15. Holland GN. Endogenous fungal infections of the retina and choroid. In: Ryan SJ, Schachat AP, Murphy RP, Patz A, eds. Retina. St. Louis: C.V. Mosby, 1989;2:625–636.

16. Edwards JE Jr, Foos RY, Montgomerie JZ, Guze LB. Ocular manifestations of *Candida septicemia*: review of seventy-six cases of hematogenous *Candida* endophthalmitis. Medicine 1974;53:47–75.

17. Parke DW II, Jones DB, Gentry LO. Endogenous endophthalmitis among patients with candidemia. Ophthalmology 1982;89:789–796.

18. Griffin JR, Pettit TH, Fishman LS, Foos RY. Blood-borne *Candida* endophthalmitis: a clinical and pathologic study of 21 cases. Arch Ophthalmol 1973;89:450–456.

19. Graham E, Chignell AH, Eykyn S. *Candida* endophthalmitis: a complication of prolonged intravenous therapy and antibiotic treatment. J Infect 1986;13:167–173.

20. Palmer EA. Endogenous *Candida* endophthalmitis in infants. Am J Ophthalmol 1980;89:388–395.

21. Montgomerie JZ, Edwards JE Jr. Association of infection due to *Candida albicans* with intravenous hyperalimentation. J Infect Dis 1978;137:197–201.

22. Servant JB, Dutton GN, Ong-Tone L, Barrie T, Davey C. Candidal endophthalmitis in Glaswegian heroin addicts: report of an epidemic. Trans Ophthalmol Soc UK 1985;104:297–308.

23. Snip RC, Michels RG. Pars plana vitrectomy in the management of endogenous *Candida* endophthalmitis. Am J Ophthalmol 1976;82:699–704.

24. Demicco DD, Reichman RC, Violette EJ, Winn WC Jr. Disseminated aspergillosis presenting with endophthalmitis: a case report and a review of the literature. Cancer 1984;53:1995–2001.

25. Doft BH, Clarkson JG, Rebell G, Forster RK. Endogenous *Aspergillus* endophthalmitis in drug abusers. Arch Ophthalmol 1980;98:859–862.

26. Naidoff MA, Green WR. Endogenous *Aspergillus* endophthalmitis occurring after kidney transplant. Am J Ophthalmol 1975;79:502–509.

27. Roney P, Barr CC, Chun CH, Raff MJ. Endogenous *Aspergillus* endophthalmitis. Rev Infect Dis 1986;8:955–958.

28. Forster RK. Fungal diseases. In: Smolin G, Thoft RA, eds. The Cornea. Boston: Little, Brown, 1983;168–177.

29. Liesegang TJ, Forster RF, Spectrum of microbial keratitis in South Florida. Am J Ophthalmol 1980;90:38–47.

30. Liesegang TJ. Bacterial and fungal keratitis. In: Kaufman HE, Barron BA, McDonald MB, Waltman SR, eds. The Cornea. New York: Churchill Livingstone, 1988;248–270.

31. Zaaka KA, Foos RY, Brown WJ. Intraocular coccidioidomycosis. Surv Ophthalmol 1978;22:313–321.

32. Rodenbiker HT, Ganley JP. Review. Ocular coccidioidomycosis. Surv Ophthalmol 1980;24:263–289.

33. Brown WC, Kellenberger RE, Hudson KE. Granulomatous uveitis associated with disseminated coccidioidomycosis. Am J Ophthalmol 1958;45:102.

34. Glasgow BJ, Brown HH, Foos RY. Miliary retinitis in coccidioidomycosis. Am J Ophthalmol 1987;104:24–27.

35. Pettit TH, Learn RN, Foos RY. Intraocular coccidioidomycosis. Arch Ophthalmol 1967;77:655–661.

36. Mandell DB, Levy JJ, Rosenthal DL. Preparation and cytologic evaluation of intraocular fluids. Acta Cytol 1987;31:150–158.

37. Shields JA, Wright DM, Augsburger JJ, Wolkowicz MI. Cryptococcal chorioretinitis. Am J Ophthalmol 1980;89:210–218.

38. Avendano J, Tanishima T, Kuwabara T. Ocular cryptococcosis. Am J Ophthalmol 1978;86:110–113.

39. Hiles DA, Font RL. Bilateral intraocular cyptococcosis with unilateral spontaneous regression: report of a case and review of the literature. Am J Ophthalmol 1968; 65:98–108.

40. Henderly DE, Liggett PE, Rao NA. Cryptococcal chorioretinitis and endophthalmitis. Retina 1987;7:75–79.

41. Kapoor S, Sood GC, Aurora AL, Sood M. Ocular cysticercosis: report of a free floating cysticercus in the anterior chamber. Acta Ophthalmol 1977;55:927–930.

42. Mathur RN, Abraham L. Cysticercosis of the eye: a case of a plastic iridocyclitis due to cysticercus cyst in the anterior chamber. Arch Ophthalmol 1962;67:562–563.

43. Kapoor S, Kapoor MS. Ocular cysticercosis. J Pediatr Ophthalmol Strabismus 1978;15:170–172.

44. Danis P. Intraocular cysticercus. Arch Ophthalmol 1974;91:238–239.

45. Topilow HW, Yimoyines DJ, Freeman HM, Young GAM, Addison R. Bilateral multifocal intraocular cysticercosis. Ophthalmology 1981;88:1166–1172.

46. Bartholomew RS. Subretinal cysticercosis. Am J Ophthalmol 1975;79:670–673.

47. Gupta AK, Aman A, Malik KPS, Arora B. Subconjunctival cysticercosis. J Pediatr Ophthalmol Strabismus 1978;15:323–325.

48. Hutton WL, Vaiser A, Snyder WB. Pars plana vitrectomy for removal of intravitreal cysticercus. Am J Ophthalmol 1976;81:571–573.

49. Santos R, Chavarria M, Aguirre AE. Failure of medical treatment in two cases of intraocular cysticercosis. Am J Ophthalmol 1984;97:249–250.

50. Wood TR, Binder PS. Intravitreal and intracameral cysticercosis. Ann Ophthalmol 1979;11:1033–1036.

51. Cano MR. Ocular cysticercosis. In: Ryan SJ. Schachat AP, Murphy RP, Patz A, eds. Retina. St Louis: C.V. Mosby, 1989;2:583–587.

52. Friedman AH, Pokorny KS, Suhan J, Ritch R, Zinn KM. Electron microscopic observations of intravitreal *Cysticercus cellulosae (Taenia solium)*. Ophthalmologica 1980;180: 267–273.

53. Kumar ND, Misra K. Fine-needle aspiration cytology of subcutaneous cysticercosis. Diagn Cytopathol 1991; 7:223–224.

54. Egbert PR, Pollard RB, Gallagher JG, Merigan TC. Cytomegalovirus retinitis in immunosuppressed hosts.II. Ocular manifestations. Ann Intern Med 1980;93:664–670.

55. Holland GN, Pepose JS, Pettit TH, Gottlieb MS, Yee RD, et al. Acquired immune deficiency syndrome: ocular manifestations. Ophthalmology 1983;90:859–873.

56. Pepose JS, Holland GN, Nestor MS, Cochran AJ, Foos RY. Acquired immune deficiency syndrome: pathogenic mechanisms of ocular disease. Ophthalmology 1985;92:472–484.

57. Pepose JS. Cytomegalovirus infections of the retina. In: Ryan SJ, Schachat AP, Murphy RP, Patz A, eds. Retina. St. Louis: C.V. Mosby, 1989;2:589–596.

58. Freeman WR, Henderly DE, Wan WL, Causey D, Trousdale MD, et al. Prevalence pathophysiology and treatment of rhegmatogenous retinal detachment in treated cytomegalovirus retinitis. Am J Ophthalmol 1987;103:527–536.

59. Urayama A, Yamada N, Sasaki T, Nishiyama Y, Watanabe H, et al. Unilateral acute uveitis with retinal periarteritis and detachment. Jpn J Clin Ophthalmol 1971;25: 607–619.

60. Culbertson WW, Blumenkranz MS, Pepose JS, Stewart JA, Curtin VT. Varicella zoster virus is a cause of the acute retinal necrosis syndrome. Ophthalmology 1986;93:559–569.

61. Freeman WR, Thomas EL, Rao NA, Pepose JS, Trousdale MD, et al. Demonstration of herpes group virus in acute retinal necrosis syndrome. Am J Ophthalmol 1986; 102:701–709.

62. Ludwig IH, Zegarra H, Zakov ZN. The acute retinal necrosis syndrome: possible herpes simplex retinitis. Ophthalmology 1984;91:1659–1664.

63. Peyman GA, Goldberg MF, Uninsky E, Tessler H, Pulido J, et al. Vitrectomy and intravitreal antiviral drug therapy in acute retinal necrosis syndrome: report of two cases. Arch Ophthalmol 1984;102:1618–1621.

64. Matsuo T, Date S, Tsuji T, Koyama M, Nakayama T, et al. Immune complex containing herpes virus antigen in a patient with acute retinal necrosis. Am J Ophthalmol 1986; 101:368–371.

65. Reese L, Sheu MM, Lee F, Kaplan JH, Nahmias A. Intraocular antibody production suggests herpes zoster is only one cause of acute retinal necrosis (ARN). Invest Ophthalmol Vis Sci 1986;27(Suppl):12.

66. Duker JS, Nielson JC, Eagle RC, Bosley TM, Granadier R, et al. Rapidly progressive acute retinal necrosis secondary to herpes simplex virus, type 1. Ophthalmology 1990;97:1638–1643.

67. Pepose JS. Acute retinal necrosis syndrome. In: Ryan SJ, Schachat AP, Murphy RP, Patz A, eds. Retina. St. Louis: C.V. Mosby, 1989;2:617–623.

68. Fisher JP, Lewis ML, Blumenkranz M. Culbertson WW, Flynn HW Jr, et al. The acute retinal necrosis syndrome. I. Clinical manifestations. Ophthalmology 1982;89: 1309–1316.

69. Nicholson DH, Wolchok EB. Ocular toxoplasmosis in an adult receiving long-term corticosteroid therapy. Arch Ophthalmol 1976;94:248–254.

70. Hoerni B, Vallat M, Durand M, Pesme D. Ocular toxoplasmosis and Hodgkin's disease. Report of two cases. Arch Ophthalmol 1978;96:62.

71. Yeo JH, Jakobiec FA, Iwamoto T, Richard G, Kreissig I. Opportunistic toxoplasmic retinochoroiditis following chemotherapy for systemic lymphoma. A light and electron microscopic study. Ophthalmology 1983;90:885.

72. Holland GN, Engstrom RE, Glasgow BJ, Berger BB, Daniels SA, et al. Ocular toxoplasmosis in patients with the acquired immunodeficiency syndrome. Am J Ophthalmol 1988; 106:653–667.

73. Rao NA, Font RL. Toxoplasmic retinochoroiditis: electron-microscopic and immunofluorescence studies of formalin-fixed tissue. Arch Ophthalmol 1977;95:273–277.

CHAPTER

9

Malignant Neoplasms of the Eye

In this chapter, malignant neoplasms of the eye are illustrated. Iris and ciliary body melanomas were presented in Chapter 5. Orbital and periorbital tumors are discussed in Chapter 10.

Only a few malignant tumors originate in the eye. Interpretation of aspirates from these tumors is not difficult if sampling is adequate and one is familiar with their specific criteria. Some primary ocular tumors have different criteria for diagnosis than analogous systemic neoplasms (e.g., melanoma). Diagnosis of other tumors is complicated by inherent difficulty in obtaining adequate material (e.g., lymphoma).

MALIGNANT LYMPHOMA

Intraocular lymphomas are rare and usually associated with involvement of the central nervous system.[1–5] Patients frequently have blurred vision without pain and characteristic yellow lesions under the retinal pigment epithelium (Figure 9–1).[6] However, ocular involvement may be confined to the vitreous cavity. When vitreous cells are present, the diagnosis may be made by vitreous aspirate or vitrectomy. Cytospin preparation or direct smears of vitreous reveal atypical lymphoid cells with prominent nucleoli (Figure 9–2). Most intraocular lymphomas contain large cells that have B-cell immunophenotypic markers. Because of the scant material obtained, it is often very difficult to make an unequivocal diagnosis of malignancy. In addition, hyaluronic acid in the vitreous may nonspecifically bind antibody making immunophenotypic markers difficult to interpret. Some authors suggest that morphologic analysis is more accurate than immunophenotypic analysis (B-and T-cell markers) in this setting.[7,8] Hyaluronidase-treated vitreous samples may improve the accuracy of immunophenotypic analysis by flow cytometry.[9]

RETINOBLASTOMA

Retinoblastoma manifests early in childhood as leukocoria.[10,11] In the genetic form it is inherited in an autosomal dominant fashion, but at the molecular level it behaves as an autosomal recessive defect.[12–16] Retinoblastoma is associated with the homozygous deletion or alteration of a well-characterized gene in the chromosome region 13q14.[17–20]

Retinoblastoma appears as a white mass arising from the retina (Figures 9–3 and 9–4). It is a small-cell tumor and is characterized by areas of necrosis, calcification, Flexner-Wintersteiner rosettes, Homer Wright rosettes, and, rarely, fleurettes (Figure 9–5).[21–24] Intraocular and orbital fine needle aspiration of retinoblastoma have been described.[25] However, it is generally recommended that fine needle aspiration and other procedures to sample retinoblastoma be avoided because seeding has been documented.[26,27] This is a controversial subject. Cytologic preparations reveal a small cell tumor with necrosis and streaming of deoxyribonucleic acid (DNA) (Figure 9–6).[28] Rosettes have been reported, but they are rarely identified.[29]

LEUKEMIA

Leukemia may involve the eye in some form in as many as 80% to 90% of cases.[30–32] Usually, it takes the form of focal infiltration of the choroid.[33] Iris involvement by leukemic cells can lead to a pseudohypopyon (Figure 9–7).[34]

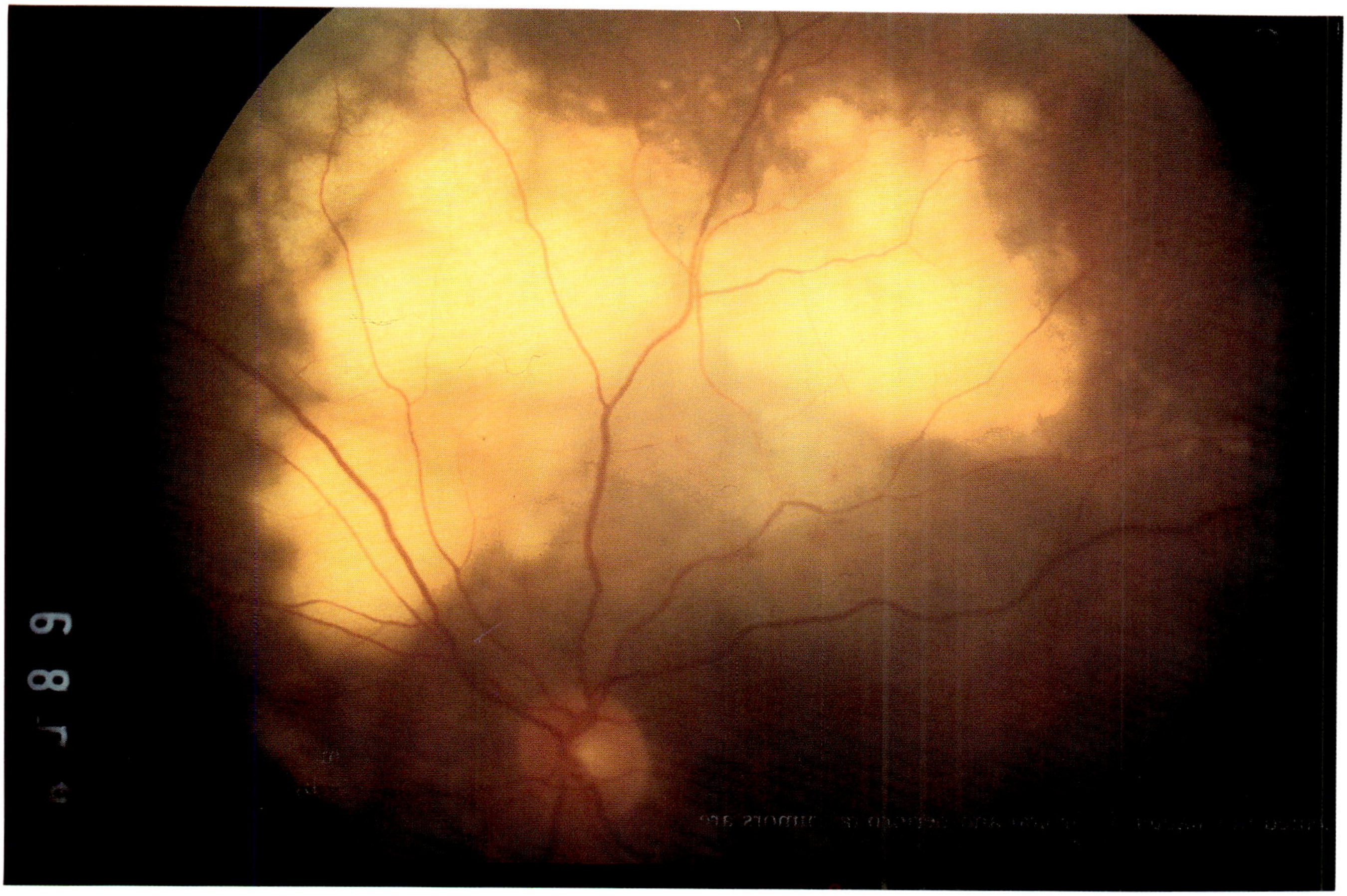

FIGURE 9–1 A 62-year-old female had blurred vision in both eyes. Clinical photographs show large yellow lesions that on fluorescein angiography had characteristics of pigment epithelial detachments with underlying infiltrates.

This may occur with acute lymphocytic leukemia, chronic lymphocytic leukemia, or acute myelogenous leukemia.[35–39] A simple aqueous aspirate with a small gauge needle is often sufficient for diagnosis (Figure 9–8).

CHOROIDAL MELANOMA

Choroidal melanoma accounts for about 80% to 87% of all ocular melanomas.[40] Patients most frequently present clinically with blurred vision and examination reveals a mass in the posterior globe (Figure 9–9). Malignant melanomas of the choroid differ from those of the skin. They have a more bland histologic appearance and they have a better prognosis (Figure 9–10).[41] Historically, choroidal melanomas are classified according to the presence of spindle or epithelioid cells.[42–44] However, it is apparent that other cytologic information may be important, including the standard deviation of the nucleolar area.[45,46] Fine needle aspiration of uveal tumors has been an effective means of diagnosis.[47–52] Criteria for cytologic diagnosis of malignant melanoma include clusters or single pigmented cells with enlarged nuclei and prominent nucleoli. Spindle cells retain their elongated shape and have nuclear grooves (Figure 9–11).[29] Cells with dendritic cytoplasm are frequently present (Figure 9–12). Epithe-

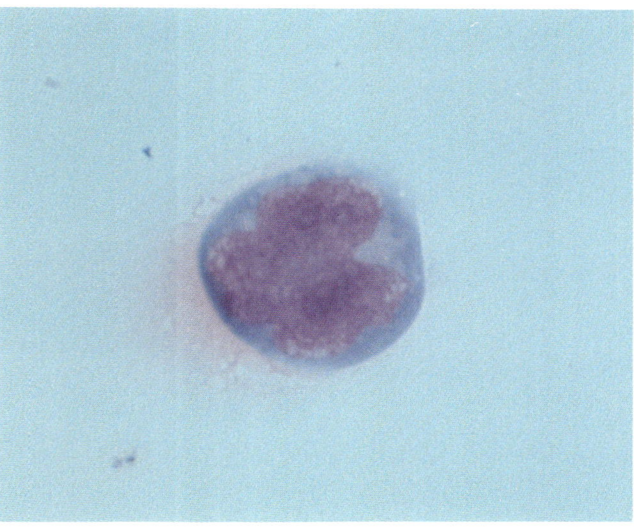

FIGURE 9–2 Direct vitreous smear from the patient in Figure 9–1 shows enlarged atypical lymphoid cells with nucleoli.

FIGURE 9–3 Gross photograph of an autopsy eye of a patient who had lymphoma in the brain and vitreous cavity. The opaque vitreous contains white infiltrate that histologically was lymphoma.

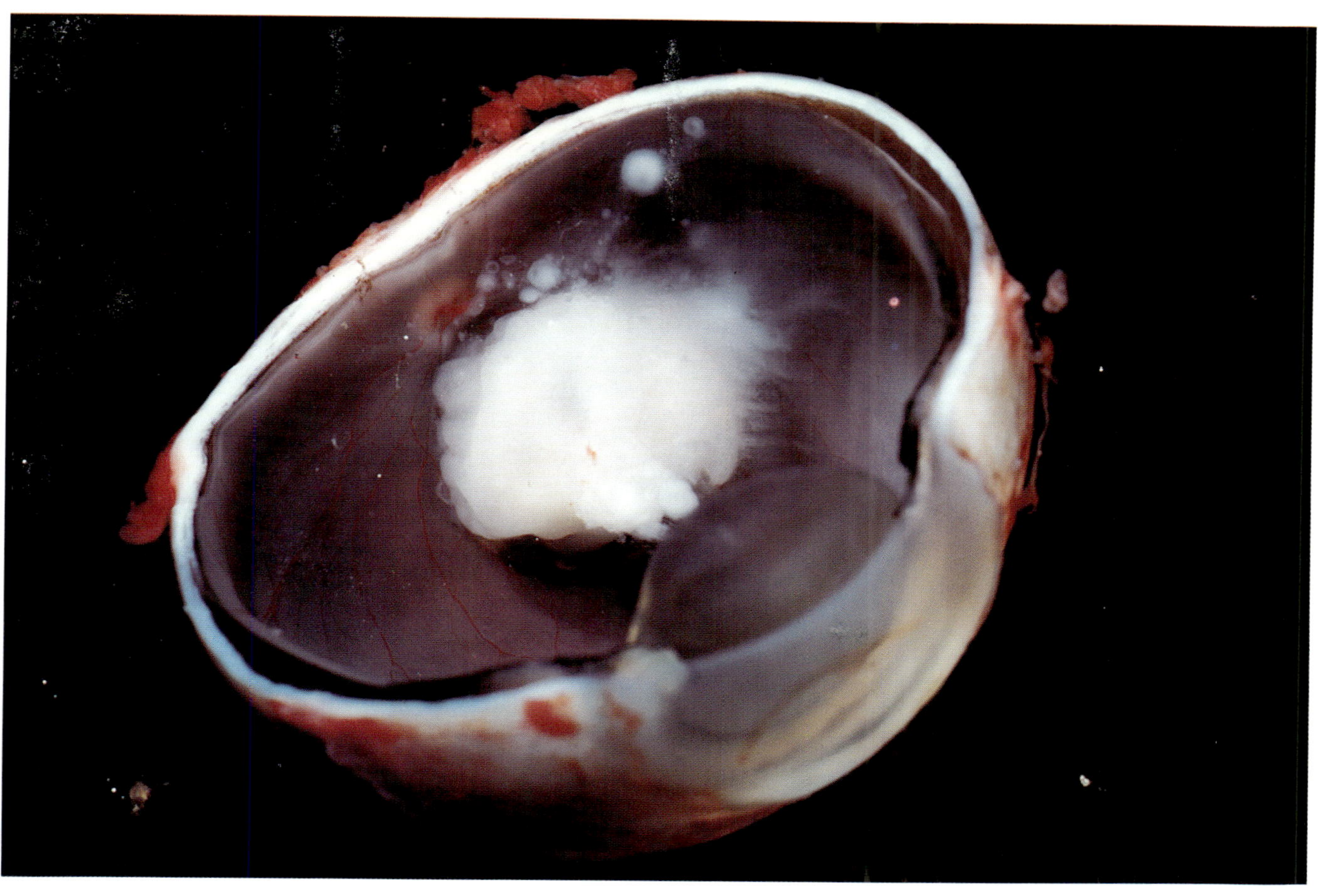

FIGURE 9–4 Gross photograph shows a white mass (retinoblastoma) arising from the retina with foci of calcification. In: Coulson WF, ed, Surgical Pathology. Philadelphia: J. B. Lippincott Co. 1988;1413.

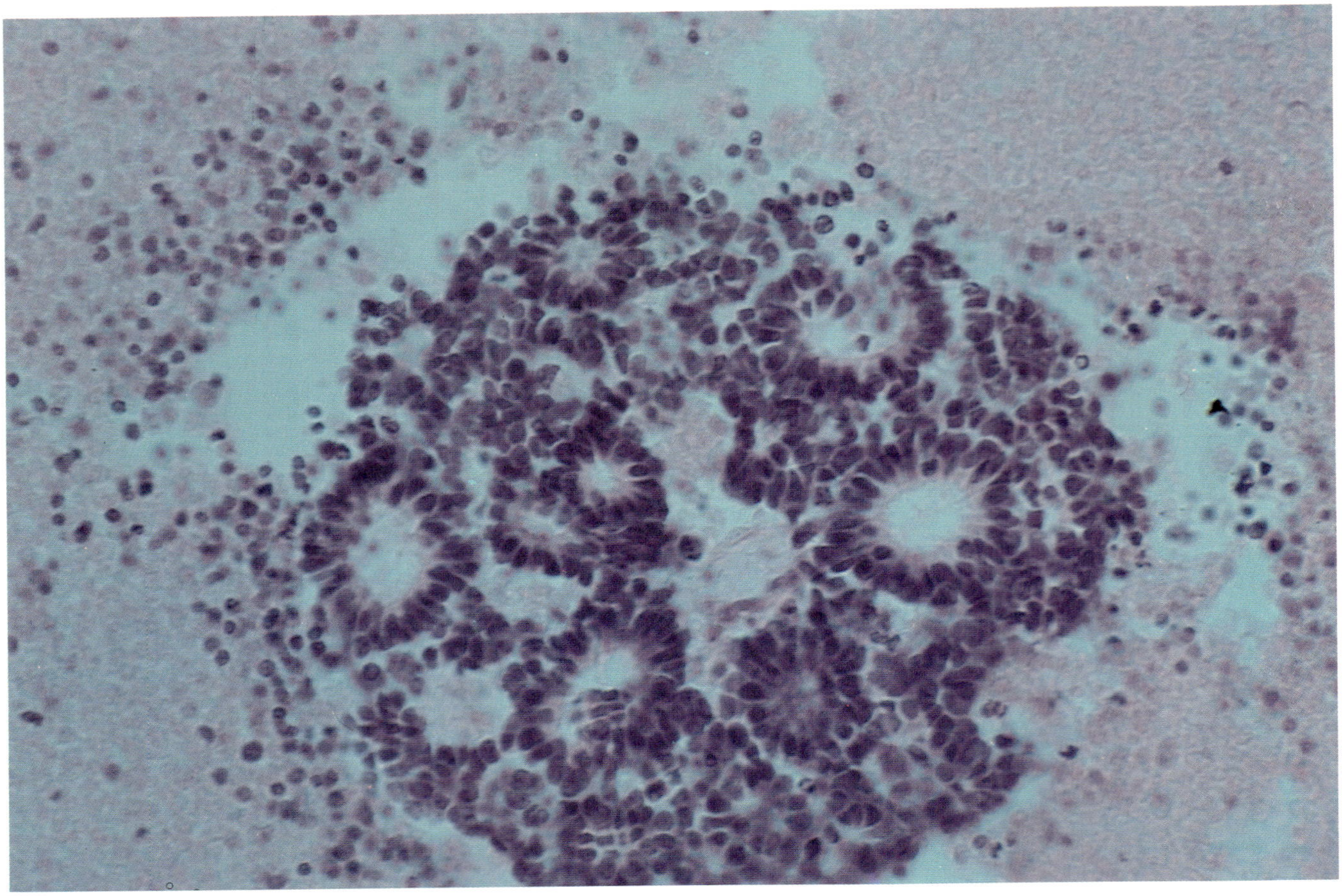

FIGURE 9–5 (ABOVE) Photomicrograph shows a small-cell tumor with foci of calcification, necrosis, and Flexner-Wintersteiner rosettes. The Flexner-Wintersteiner rosette is comprised of retinoblastoma cells radially arranged around a lumen lined by a delicate limiting membrane.

FIGURE 9–6 (RIGHT) Photomicrograph shows numerous small cells with scant cytoplasm. Many of the cells are disrupted and DNA is smeared on the slide. (May-Grünwald Giemsa, × 400)

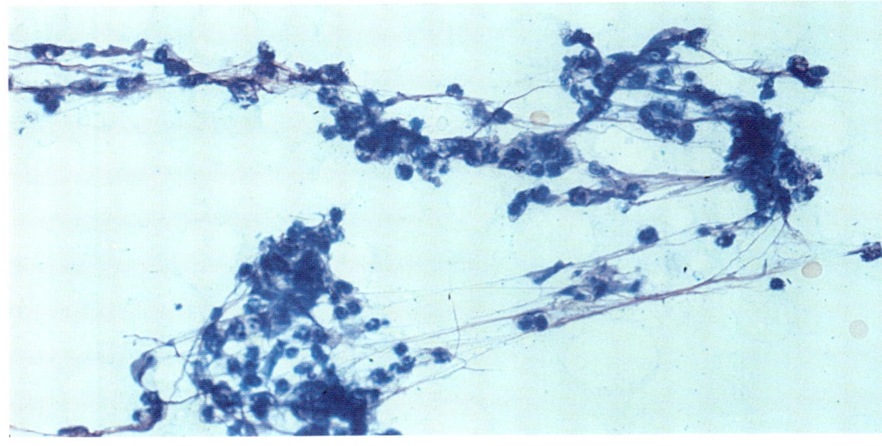

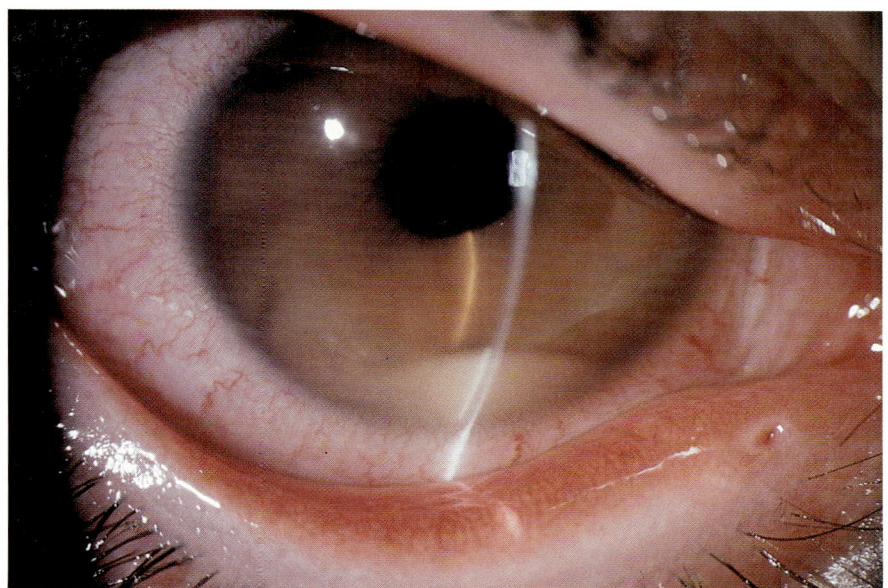

FIGURE 9–7 Clinical photograph of a pseudohypopyon. The patient had been successfully treated for acute lymphocytic leukemia. The initial manifestation of recurrence was this white-yellow layered meniscus in the anterior chamber. Courtesy of Associate Professor Gary N. Holland, M.D., Jules Stein Eye Institute.

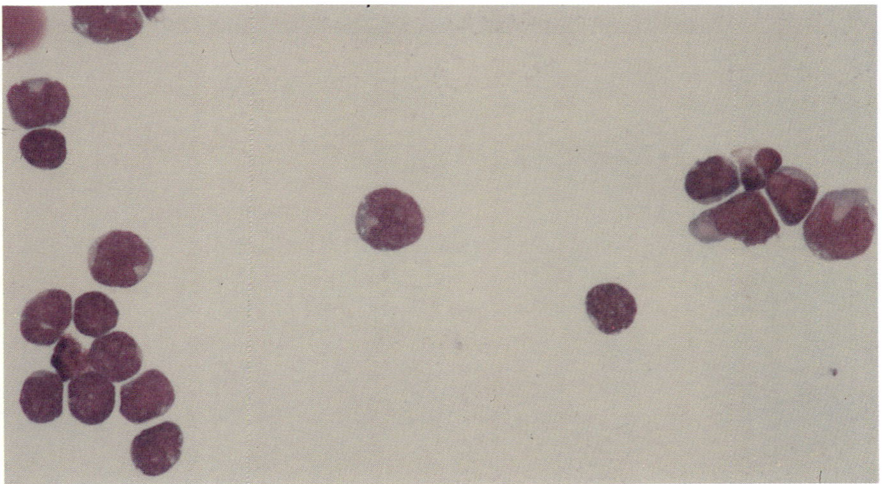

FIGURE 9–8 Anterior chamber aspirate from the patient in Figure 9–7 shows numerous single atypical round cells with enlarged nuclei and prominent nucleoli (lymphoblasts). (May-Grünwald Giemas, × 720)

FIGURE 9–9 Gross photograph shows a dome-shaped lightly pigmented tumor (melanoma) and an associated exudative retinal detachment.

lioid cells have more abundant cytoplasm, larger round nuclei, and more prominent nucleoli (Figure 9–13). Occasionally, enlarged cells with vacuolated cytoplasm (balloon cells) are sampled (Figure 9–14). Invariably present in fine needle aspirate smears are macrophages that are stuffed with pigment (Figure 9–15).

METASTATIC CARCINOMA

One source of difficulty in the differential diagnosis of ocular tumors is metastatic disease. The incidence of metastatic ocular cancer in autopsy studies varies from 0.06% to 2.3%.[53,54] The most common site for metastasis in the eye is the choroid.[55,56] The site of the primary lesion is most often the breast (46% to 90%) followed by the lung (10% to 29%).[57] Because metastatic lesions may produce dome-shaped tumors in the choroid, fine needle aspiration may be necessary to differentiate a primary ocular amelanotic melanoma from a metastatic cancer (Figure 9–16). The cytologic findings of metastatic tumors correspond to those of the primary tumor and occasionally, the features may suggest a primary site (Figure 9–17).

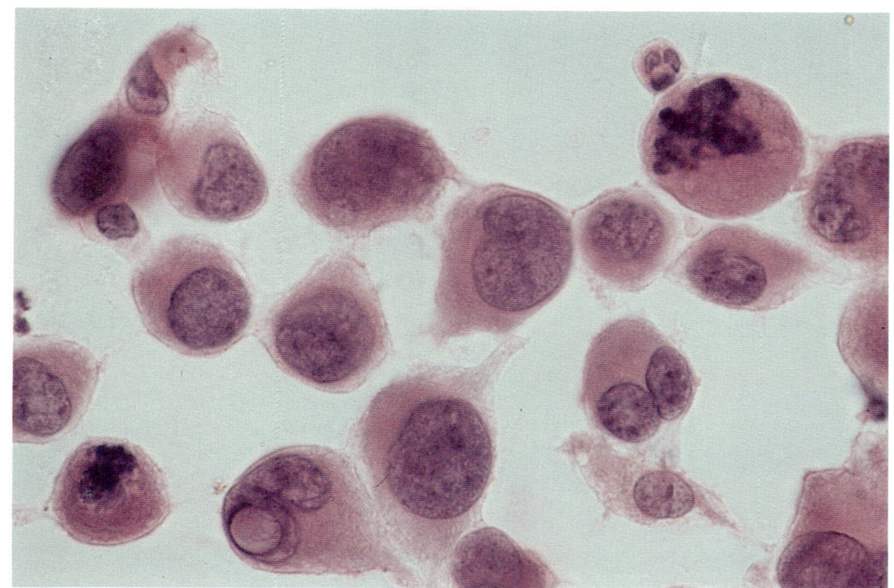

FIGURE 9–10 Photomicrograph of a fine needle aspirate from a metastatic melanoma. Compare the very large anaplastic cells with those shown in Figures 9–11 and 9–13. (hematoxylin and eosin, × 500)

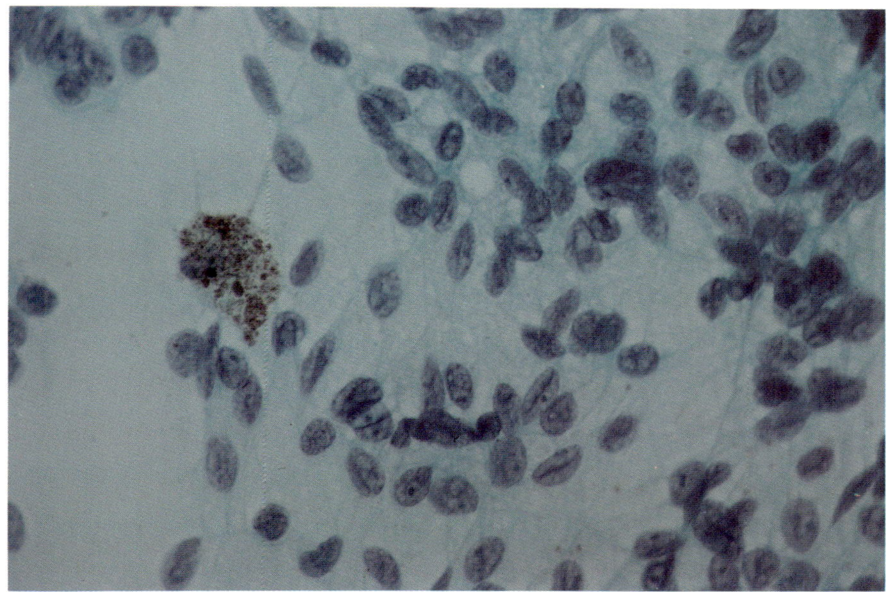

FIGURE 9–11 Aspirate from an ocular melanoma demonstrates spindle cells containing brown pigment with elongated nuclei, nucleoli, and nuclear grooves parallel to the long axis of the nuclei. (Papanicolaou, × 250)

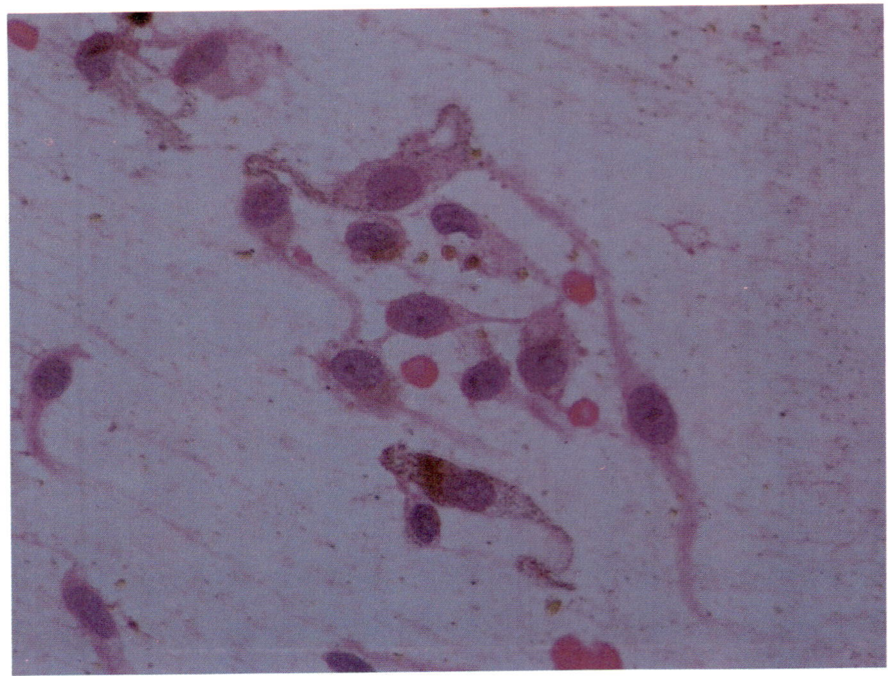

FIGURE 9–12 Photomicrograph from a mixed-cell melanoma shows dendritic cytoplasm, abundant cytoplasm with brown pigment.

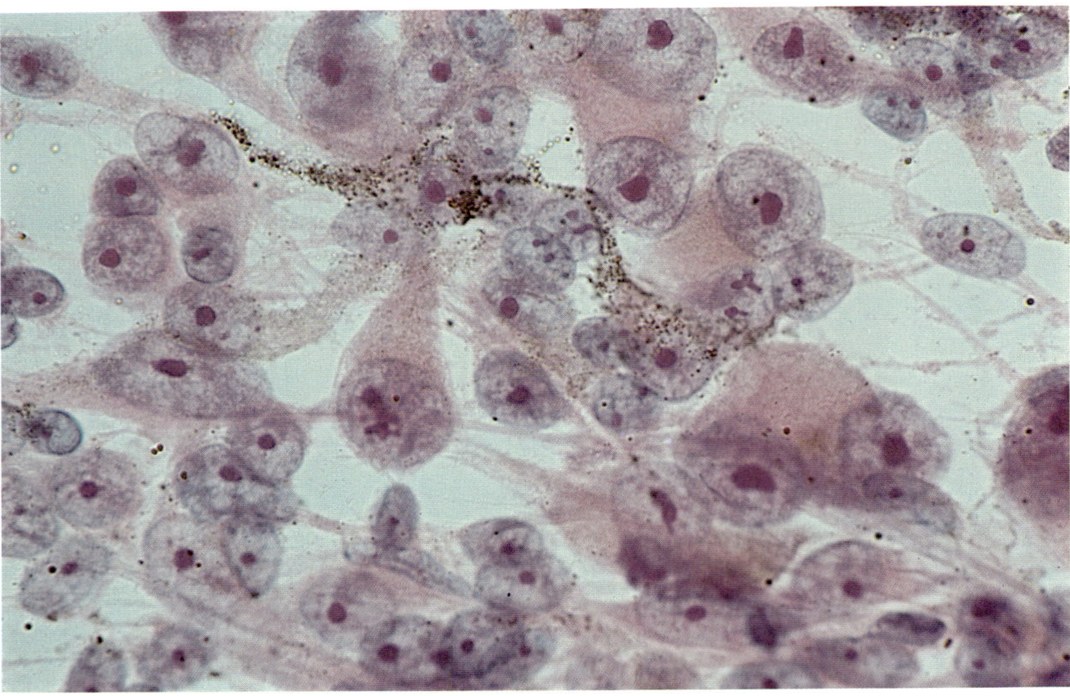

FIGURE 9–13 Photomicrograph shows epithelioid melanoma cells with abundant cytoplasm, pigment within the cytoplasm, and prominent nucleoli.

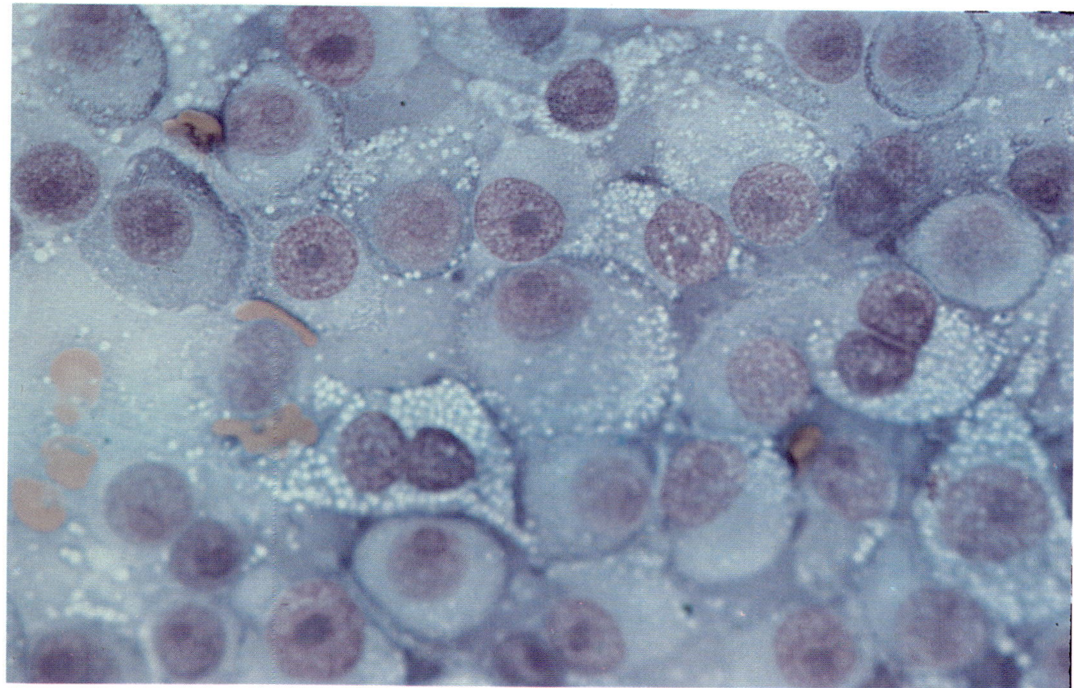

FIGURE 9–14 **Photomicrograph shows a lipidized melanocyte, balloon cell from a fine needle aspirate of an ocular melanoma. (hematoxylin and eosin, × 512)**

REFERENCES

1. Barr CC, Green WR, Payne JW, Knox DL, Jensen AD, et al. Reticulum cell sarcoma of the retinal and uvea. Surv Ophthalmol 1975;19:224–239.
2. Parver LM, Font RL. Malignant lymphoma of the retina and brain. Arch Ophthalmol 1979;97:1505–1507.
3. Wagoner MD, Gonder JR, Albert DM, Canny CL. Intraocular reticulum cell sarcoma. Ophthalmology 1980;87:724–727.
4. Lang GK, Surer JL, Green WR, Finkelstein D, Michels RG, et al. Ocular reticulum cell sarcoma. Retina 1985;5:79–85.
5. Michelson JB, Michelson PE, Bordin GM, Chisari FV. Ocular reticulum cell sarcoma. Arch Ophthalmol 1981;99: 1409–1411.
6. Qualman SJ, Mendelsohn G, Mann RB, Green WR. Intraocular lymphomas. Cancer 1983;52:878–886.
7. Char DH, Ljung BM, Deschenes J, Miller TR. Intraocular lymphoma: immunological and cytological analysis. BR J Ophthalmol 1988;72:905–911.
8. Ljung BM, Char D, Miller TR, Deschenes J. Intraocular lymphoma, cytologic diagnosis and the role of immunologic markers. Acta Cytol 1988;32:840–847.
9. Wilson DJ, Braziel R, Rosenbaum JT. Intraocular lymphoma: Immunopathologic analysis of vitreous biopsies. Invest Ophthalmol 1990;31:368.
10. Albert DM. Historic review of retinoblastoma. Ophthalmology 1987;94:654.

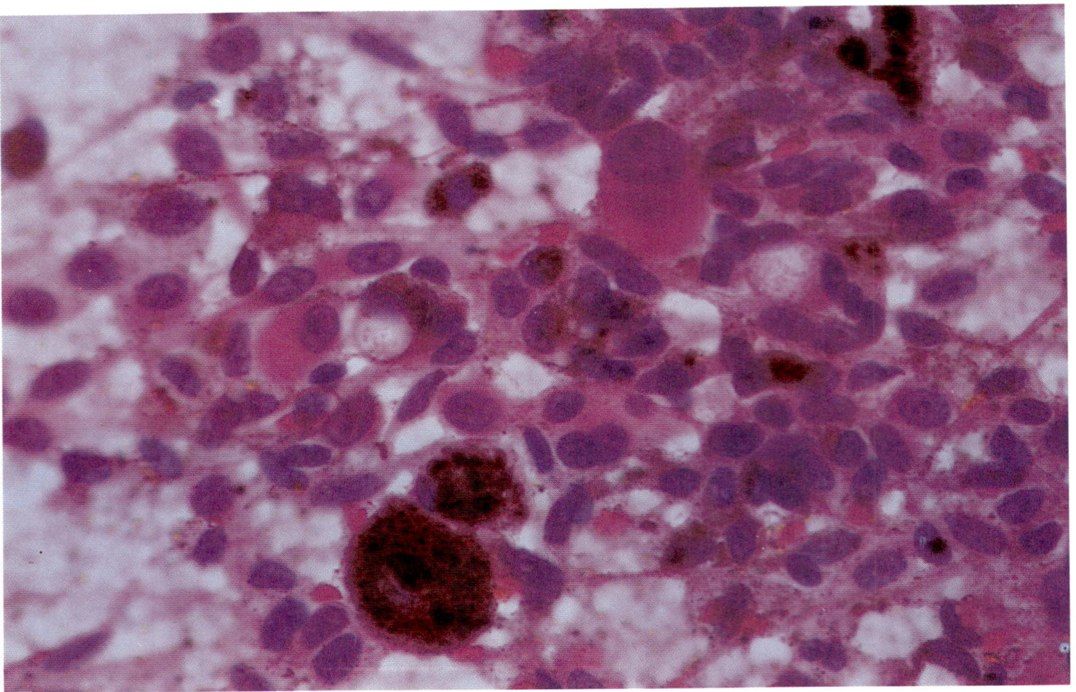

FIGURE 9–15 **Photomicrograph shows two cells with cleaved small nuclei and numerous large pigmented granules in the abundant cytoplasm. Contrast these macrophages to adjacent tumor cells with prominent nucleoli, enlarged nuclei, and more finely granular cytoplasmic pigment. (hematoxylin and eosin, × 512)**

11. Sang DN, Albert DM. Retinoblastoma: clinical and histopathologic features. Hum Pathol 1982;13:133–147.

12. Dryja TP, Cavenee W, White R, Rapaport JM, Peterson R, et al. Homozygosity of chromosome qe in retinoblastoma. N Engl J Med 1984;310:550.

13. Vogel F. Genetics of retinoblastoma. Hum Genet 1979;52:1–54.

14. Murphree AL, Benedict WF. Retinoblastoma: clues to human oncogenesis. Science 1984;223:1028.

15. Wiggs DL, Dryja TP. Predicting the risk of hereditary retinoblastoma. Am J Ophthalmol 1988;106:346–351.

16. Dunn JM, Phillips RA, Becker AJ, et al. Identification of germline and somatic mutations affecting the retinoblastoma gene. Science 1988;241:1797.

17. Lee W-H, Bookstein R, Hong F, Young L-J, Shew J-Y, et al. Human retinoblastoma susceptibility gene: cloning, identification and sequence. Science 1987;235:1394–1399.

18. Friend SH, Bernards R, Rogelj S, Weinberg RA, Rapaport JM, et al. A human DNA segment with properties of the gene that predisposes to retinoblastoma and osteosarcoma. Nature 1986;323:643–646.

19. Harbour JW, Lai S-L, Whang-Peng J, Gazdar AF, Minna JD, et al. Abnormalities in structure and expression of the human retinoblastoma gene in SCLC. Science 1988;241: 353.

20. T'ang A, Varley JM, Chakraborty S, Murphree AL, Fung Y-KT, et al. Structural rearrangement of the retino-

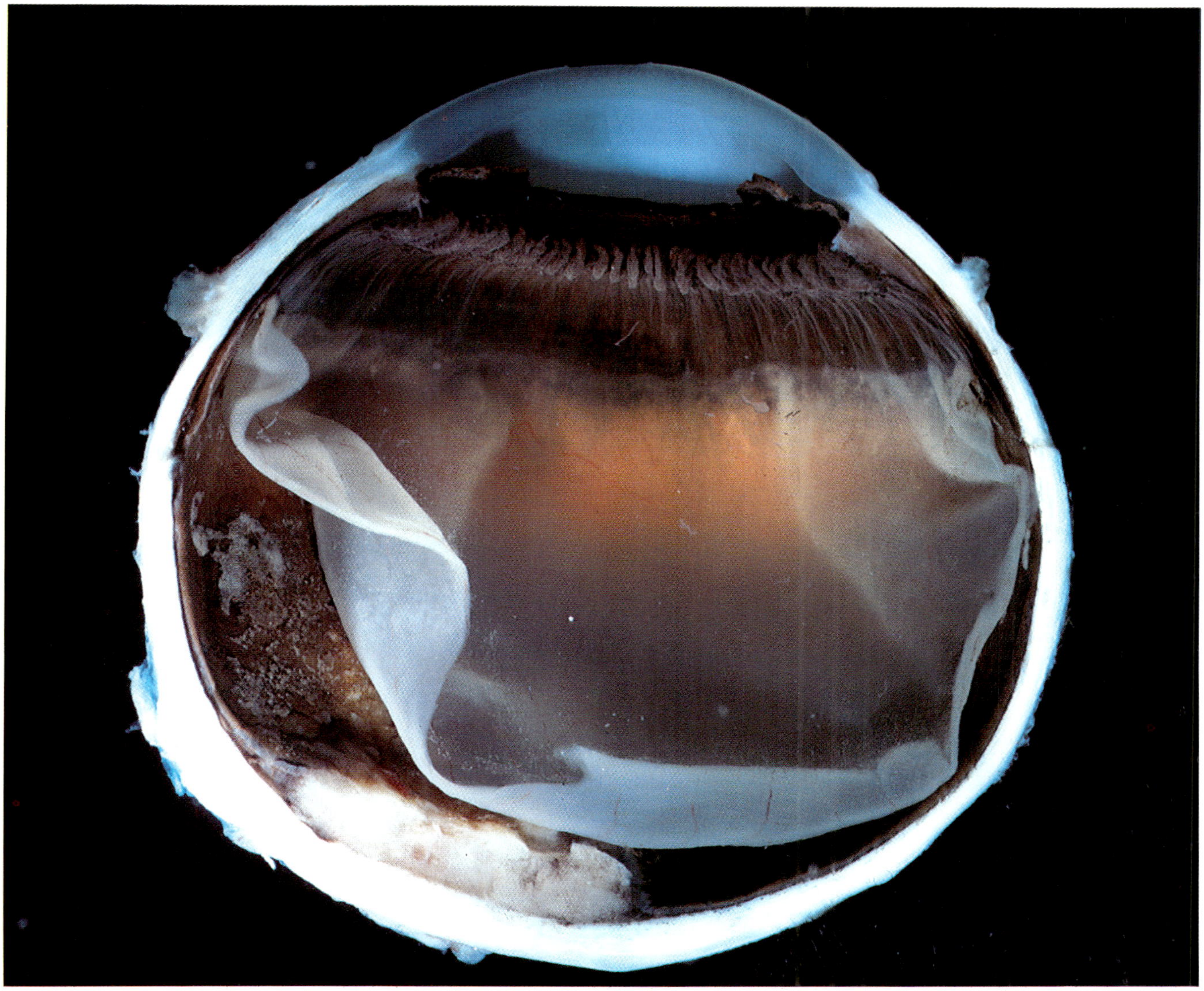

FIGURE 9–16 Gross photograph of a sectioned eye of a patient with a metastatic choroidal neoplasm. The tumor is dome shaped and contains focal areas of light pigmentation (hemosiderin).

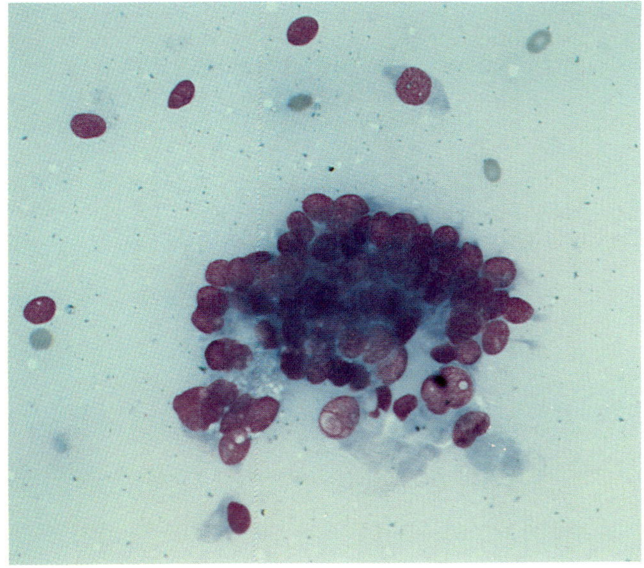

FIGURE 9–17 Aspiration of the tumor shown in Figure 9–16 shows clusters of cells arranged in a papillary architecture with enlarged nuclei and prominent inclusions within the nuclei (orphan Annie) consistent with a papillary carcinoma. The primary site was demonstrated to be the thyroid.

blastoma gene in human breast carcinoma. Science 1988; 242:263.

21. Tso MOM, Fine BS, Zimmerman LE. The nature of retinoblastoma. II. An electron microscope study. Am J Ophthalmol 1970;69:350–359.

22. Tso MOM, Fine BS, Zimmerman LE. The Flexner-Wintersteiner rosettes in retinoblastoma. Arch Pathol 1969;88:664–671.

23. Rodrigues MM, Wiggert B, Shields J, Danoso L, Bardenstein D, et al. Retinoblastoma. Immunohistochemistry and cell differentiation. Ophthalmology 1987;94:378.

24. Yanoff M, Fine BS. Ocular pathology. A text and atlas. Philadelphia: J.B. Lippincott, 1989.

25. Char DH, Miller TR. Fine needle biopsy in retinoblastoma. Am J Ophthalmol 1984;97:686–690.

26. Rodriguez A. Diagnosis of retinoblastoma by cytologic examination of the aqueous and vitreous. Mod Probl Ophthalmol 1977;18:142–148.

27. Karcioglu ZA, Gordon RA, Karcioglu GL. Tumor seeding in ocular fine needle aspiration biopsy. Ophthalmology 1985;92:1763–1767.

28. Rosenthal DL, Mandell DB, Glasgow BJ. Eye. In: Bibbo M, ed. Comprehensive cytology. Philadelphia: W.B. Saunders, 1991;484–501.

29. Scroggs MW, Johnston WW, Klintworth GK. Intraocular tumors: a cytopathologic study. Acta Cytol 1990;34:401–408.

30. Duke-Elder WS, ed. System of opthalmology. St Louis, C.V. Mosby, 1965;13:817–818.

31. Allen RA, Straatsma BR. Ocular involvement in leukemia and allied disorders. Arch Ophthalmol 1961;66:490–508.

32. Rosenthal AR. Ocular manifestations of leukemia. Ophthalmology 1983;90:899–905.

33. Kincaid MC, Green WR. Ocular and orbital involvement in leukemia. Surv Ophthalmol 1983;27:211–232.

34. Perry HD, Mallen FJ. Iris involvement in granulocytic sarcoma. Am J Ophthalmol 1979;87:530–532.

35. Martin B. Infiltration of the iris in chronic lymphatic leukemia. Br J Ophthalmol 1968;52:781–785.

36. Fonken HA, Ellis PP. Leukemic infiltrates in the iris. Arch Ophthalmol 1966;76:32-36.

37. Holbrook CT, Elsas EJ, Crist WM, et al. Acute leukemia and hypopyon. J Pediat 1978;93:626–628.

38. Johnston SS, Ware CF. Iris involvement in leukemia. Br J Ophthalmol 1973;57:320–324.

39. Kincaid MC, Green WR, Kelley JS. Acute ocular leukemia. Am J Ophthalmol 1979;87:698–702.

40. Reese AB. Tumors of the eye. 3d edition. Hagerstown, MD Harper & Row, 1976:229.

41. Zimmerman LE. Malignant melanoma of the uveal tract. In: Ophthalmic pathology. Philadelphia: W.B. Saunders Co, 1986;2072–2139.

42. Callender GR. Malignant melanotic tumors of the eye. A study of histologic types in 111 cases. Trans Am Acad Ophthalmol Otolaryngol 1931;367:131.

43. McLean IW, Foster WD, Zimmerman LE. Modifications of Callender's classification of uveal melanoma at the Armed Forces Institute of Pathology. Am J Ophthalmol 1983;96:502–509.

44. Paul EV, Parnell BL, Fraker M. Prognosis of malignant melanomas of the choroid and ciliary body. Int Ophthalmol Clin 1962;2:387.

45. Gamel JW, McLean IW, Greenberg RA, Zimmerman LE, Lichtenstein SJ. Computerized histologic assessment of malignant potential: a method for determining the prognosis of uveal melanomas. Hum Pathol 1982;13:893–897.

46. Gamel JW, McLean IW, Greenberg RA, Naids RM, Folberg R, et al. Objective assessment of the malignant potential of intraocular melanoma with standard microslides stained with hematoxylin-eosin. Hum Pathol 1985;16:689–692.

47. Jakobiec FA, Coleman DJ, Chattock A, Smith M. Ultrasonically guided needle biopsy and cytologic diagnosis of solid intraocular tumors. Ophthalmology 1979;86:1662–1678.

48. Augsburger JJ, Shields JA. Fine needle aspiration biopsy of solid intraocular tumors. Indications, instrumentation and techniques. Ophthalmic Surg 1984;15:34–40.

49. Czerniak B, Woyke S, Domagala W, Krzysztolik Z. Fine needle aspiration cytology of intraocular malignant melanoma. Acta Cytol 1983;27:157–165.

50. Midena E, Segato T, Piermarocchi S, Boccato P. Fine needle aspiration biopsy in ophthalmology. Surv Ophthalmol 1985;29:410–422.

51. Davey CC, Deery ARS. Through the eye, a needle. Intraocular fine needle aspiration biopsy. Trans Ophthalmol Soc U.K. 1986;195:78–83.

52. Char DH, Miller TR, Ljung B-M, Howes EL, Stoloff A. Fine needle aspiration biopsy in uveal melanoma. Acta Cytol 1989;33:599–605.

53. Gotfredsen E. On the frequency of secondary carcinomas in the choroid. Acta Ophthalmol 1944;22:394–400.

54. Albert DM, Rubenstein RA, Scheie HG. Tumor metastasis to the eye. I. Incidence in 213 adult patients with generalized malignancy. Am J Ophthalmol 1967;63:723–726.

55. Block RS, Gartner S. The incidence of ocular metastatic carcinoma. Arch Ophthalmol 1971;85:673–675.

56. Stephens RF, Shields JA. Diagnosis and management of cancer metastatic to the uvea. A study of 70 cases. Ophthalmology 1977;86:1336–1349.

57. Ferry AP, Font RL. Carcinoma metastatic to the eye and orbit. I. A clinicopathologic study of 227 cases. Arch Ophthalmol 1974;92:276–286.

CHAPTER

10

Fine Needle Aspiration of Orbital and Periorbital Lesions

In this chapter, cytologic findings in orbital fine needle aspiration are illustrated. The general pathologists and cytopathologists who are skilled in interpretation of extra-orbital fine needle aspirates will readily transfer that expertise to orbital aspirates because the tumors are similar. Accuracy in differentiating benign and malignant orbital lesions by aspiration cytology varies from 50% to 100% in previous series.[1,2] The utility of fine needle aspiration is determined by how the results will affect patient management. If patient management is unaffected, then the procedure is not necessary. For example, certain orbital tumors are removed completely during the first operation (mixed tumor of lacrimal gland, schwannoma, dermoid cyst, and cavernous hemangioma). If the clinical diagnosis of these lesions is certain prior to the operation, then fine needle aspiration biopsy is not required. However, the clinical and radiologic diagnoses are frequently wrong (over 50% of cases in one series) and fine needle aspiration has been useful.[3] There are two general indications in which fine needle aspiration of the orbit has proven most useful. First, if the suspected tumor can be treated without surgical intervention (e.g., rhabdomyosarcoma, sarcoidosis, metastatic cancer, reactive lymphoid hyperplasia, lymphoma, sclerosing orbititis, and infections), an accurate fine needle biopsy may spare the patient any further procedure. Second, fine needle aspiration may help the surgeon plan an operation. For example, a patient with the erroneous clinical or radiologic diagnosis of osteosarcoma might have a planned radical orbitectomy changed to an appropriate curettage if a fine needle aspiration biopsy reveals eosinophilic granuloma. In addition, the fine needle biopsy may radically change a medical evaluation for metastatic disease. The patient with a suspected lymphoma requires a different evaluation than the patient with granulomatous disease.

SMALL-CELL TUMORS

Fine needle aspiration is an excellent way to categorize small-cell tumors. However, further specification requires more extensive immunocytochemical or electron microscopic evaluation.

Rhabdomyosarcoma

Orbital rhabdomyosarcoma is the most common primary malignant orbital neoplasm in children.[4,5] It often presents as a rapidly expanding lesion in the orbit and occurs most commonly at about age 6 or 7 (Figure 10–1).[6,7] Fine needle aspiration smears show a small-cell tumor with scant cytoplasm (Figure 10–2). Rhabdomyosarcoma of the orbit has been reported to be specifically diagnosed in at least six previous cases with fine needle aspiration biopsy.[3-10] In two other cases, a cytologic diagnosis of malignancy could be made, but not further specified.[9] To maximize the diagnostic potential of orbital aspiration cytology, other techniques, such as immunocytochemistry for desmin, actin, and myosin, or electron microscopic demonstration of myofibrils, are often required (Figure 10–3).[10] If a specific diagnosis can be made, fine needle aspiration has the advantages of being rapid and avoiding open biopsy. Radiation and chemotherapy are successful in 90% of cases.[11]

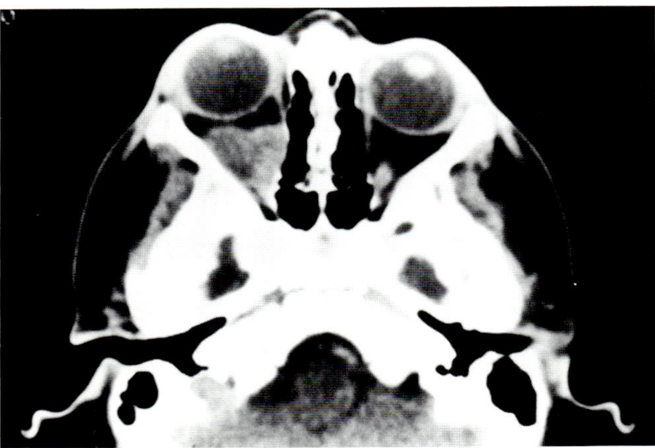

FIGURE 10–1 Computer-assisted tomogram shows an orbital mass that fills the orbit and produced proptosis in this child.

Reactive Lymphoid Hyperplasia

If one excludes basal-cell carcinomas, lymphoid lesions are the most common tumors in series of orbital fine needle aspiration.[8,9]

Benign reactive lymphoid hyperplasia is a pathologic diagnosis given to orbital and periorbital lymphoid infiltrates that demonstrate germinal centers with tingible body macrophages and a heterogenous population of cells, including small lymphocytes, reactive lymphocytes (immunoblasts), and plasma cells. This diagnosis is made by identification of the heterogenous lymphoid elements. Architecture (diffuse or nodular) cannot be discerned in fine needle aspirates (Figure 10–4). Five percent to 25% of patients with the histologic appearance of a reactive process will eventually develop systemic lymphoma.[12] Lesions intermediate between benign reactive lymphoid hyperplasia and lymphoma have been called atypical lymphoid hyperplasia. These lesions have irregular follicles, more atypical lymphocytes, and perhaps a slightly worse prognosis.[13] However, reactive lesions by morphologic and immunophenotypic criteria may harbor clones of proliferating B cells.[14,15] These clones are identified on Southern blots as nongermline bands that have the same size DNA fragments when cleaved by restriction endonucleases and hybridized to radioactively labeled probes from specific

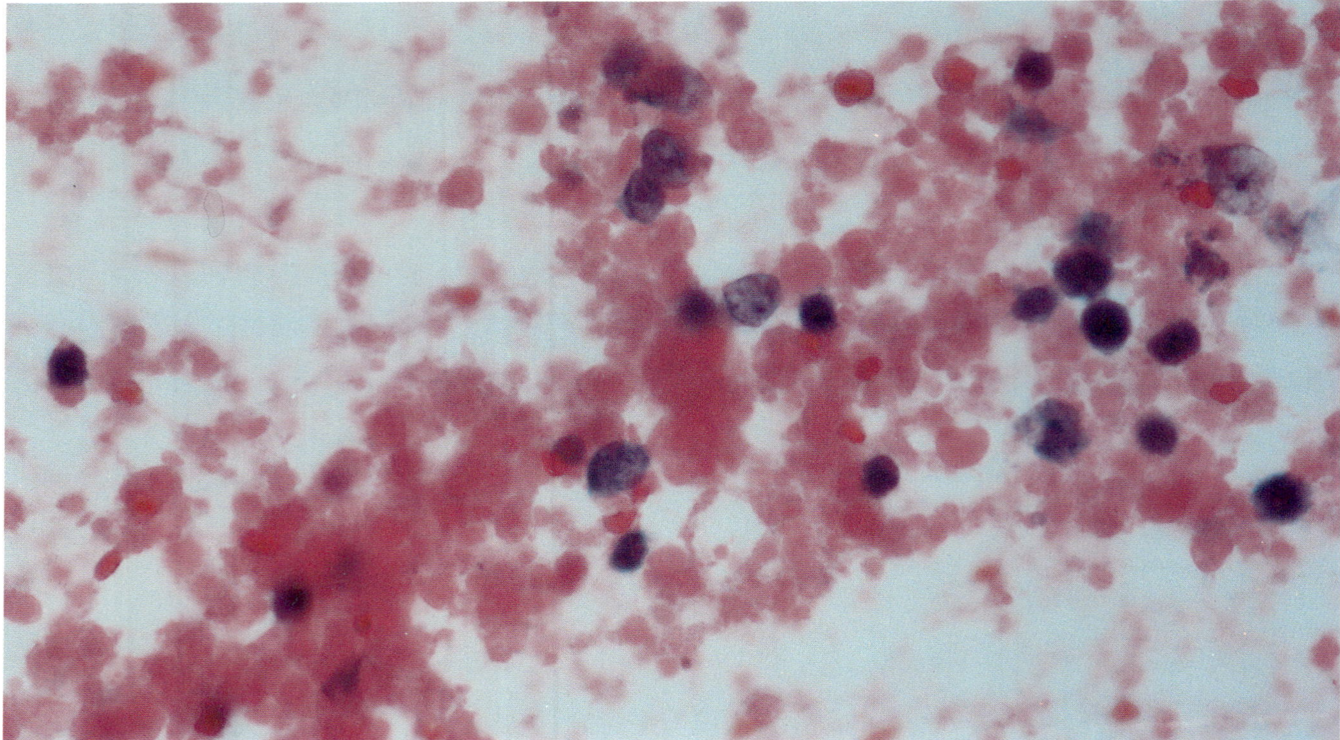

FIGURE 10–2 Aspiration cytology was performed under ultrasound with mask anesthesia. Numerous small cells with scant cytoplasm allow categorization of this lesion as a small-cell tumor. (hematoxylin and eosin, × 400)

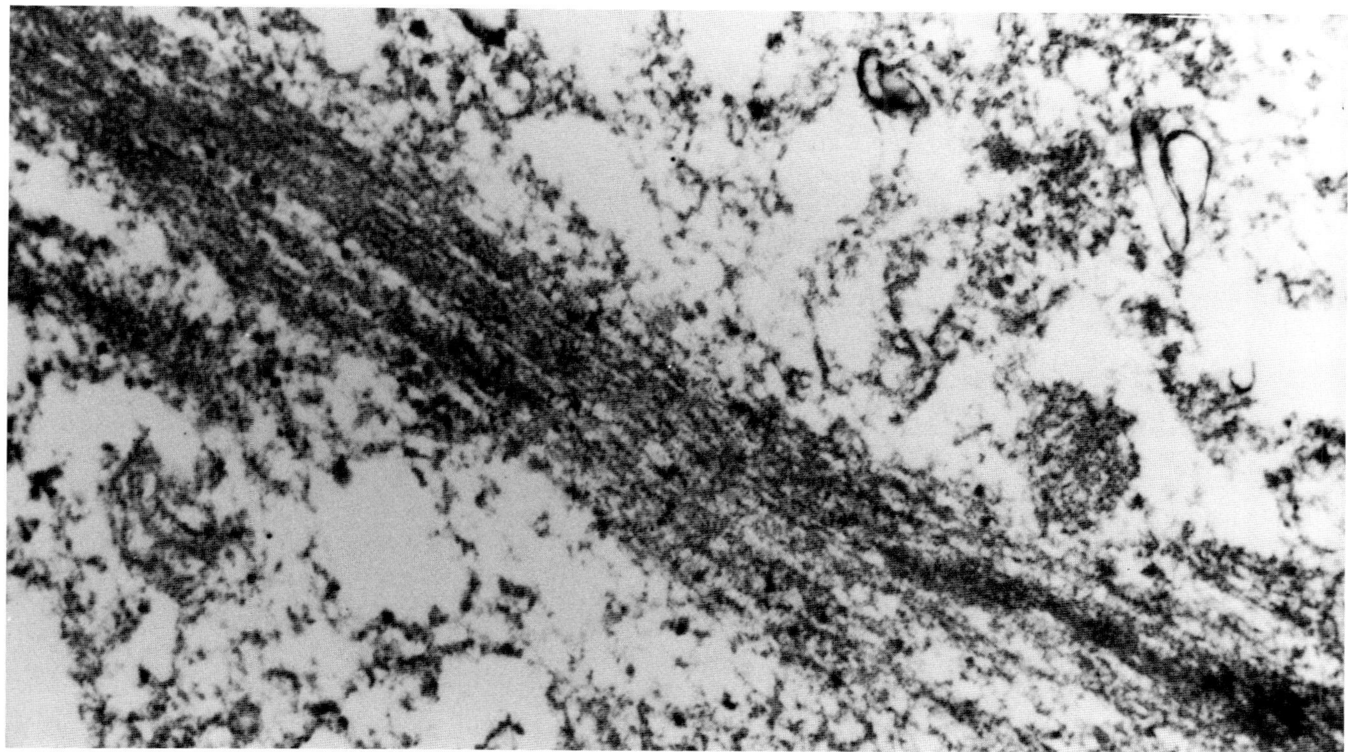

FIGURE 10–3 (ABOVE) Electron photomicrograph shows fibrils in the cytoplasm indicative of rhabdomyosarcoma. The patient was seen at another hospital and an open biopsy confirmed the fine needle aspiration diagnosis of rhabdomyosarcoma.

FIGURE 10–4 (BELOW) Photomicrograph from a fine needle aspiration shows scant cells, but a heterogenous population, including large reactive lymphocytes and small round ones. (May-Grünwald Giemsa, × 480)

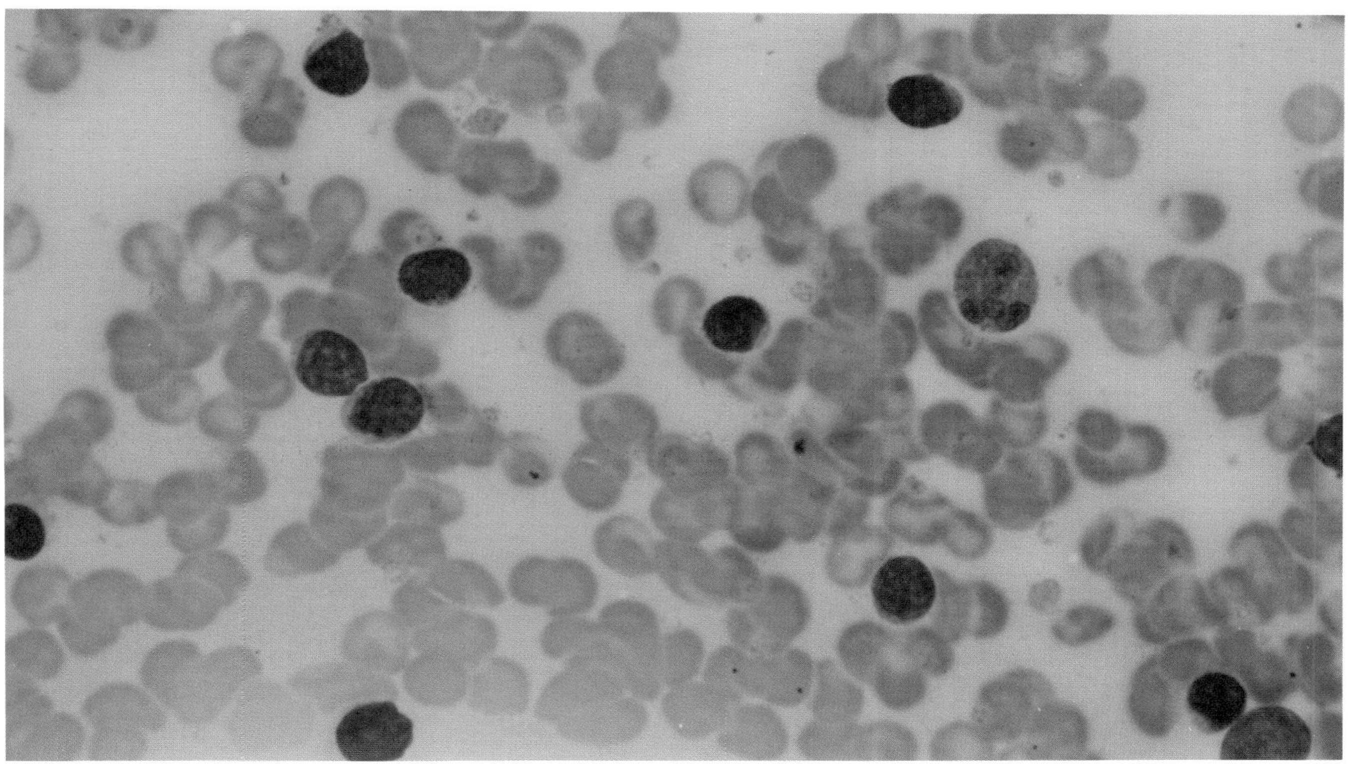

sites on the immunoglobulin gene.[16-20] It is becoming evident that orbital lymphoid lesions are a spectrum of B-cell neoplasias.[21] Because no specific morphologic, immunophenotypic, or molecular criteria have yet been well correlated with eventual outcome, it is not clear what investigative procedures are required in order to manage patients with lymphoid lesions. If treatment is to be based on clinical criteria, then fine needle aspiration biopsy with or without phenotypic markers will suffice. However, if the type of treatment is predicated on gene rearrangement studies, then at this time open biopsy with removal of adequate tissue is necessary.

Malignant Lymphoma

Patients with orbital lymphomas are usually 50 to 60 years old and present insidiously with proptosis and, frequently, a rubbery fleshy mass under the bulbar conjunctiva or eyelid (Figure 10–5). CT scan may show that the tumor conforms to the contours of the eye (Figure 10-6). Fine needle aspiration reveals abundant cellular material with a relatively homogenous population of lymphocytes (Figure 10–7). Orbital lymphomas are similar to other extranodal non-Hodgkin lymphomas. They are usually diffuse and of B-cell lineage.

Although histologic patterns of diffuse and nodular lymphoma cannot be discerned from aspiration smears, B- and T-cell marker studies can be done (Figure 10–8).[22] The prognosis of orbital lymphomas may be better determined by location and involvement of other sites than histologic classification and marker studies.[23,24]

Idiopathic Orbital Inflammation (Sclerosing Orbititis)

Idiopathic orbital inflammation is also referred to as inflammatory pseudotumor. It may present acutely or in a chronic form. In the acute form, there is an abrupt onset of pain, injection, chemosis, and decreased ocular motility. The inflammation involves orbital soft tissues, including fat, extraocular muscle, tendon, lacrimal gland, and blood vessels.[25,26] Sclera may be inflamed in late cases.[27] In the chronic form, there is marked fibrosis that may envelop the structures of the orbit and mimic a malignant neoplasm. The diagnosis often can be made clinically, but occasionally a fine needle aspiration biopsy will be requested. It is extremely difficult to get adequate material to diagnose chronic sclerosing orbititis. Usually, scant aspirates with a few fibroblasts and lymphocytes will be identified (Figure 10–9).

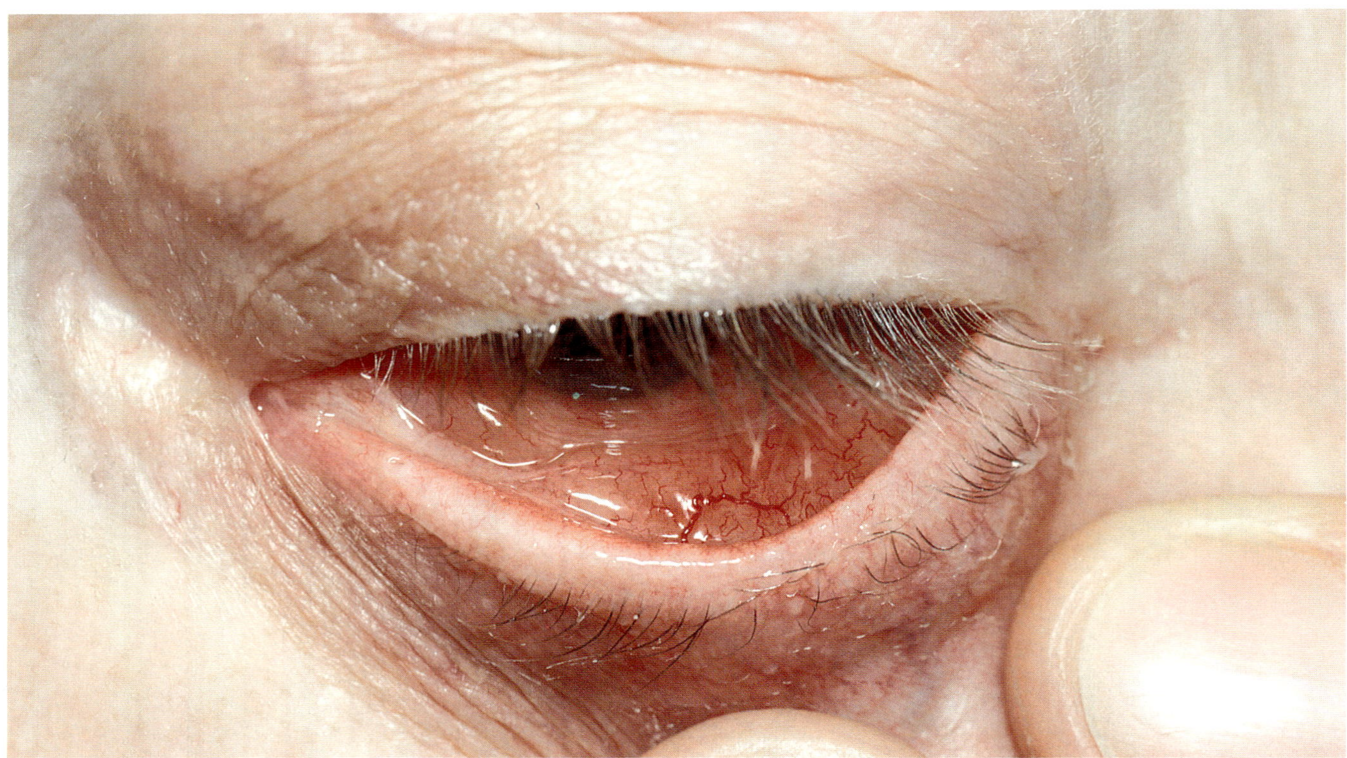

FIGURE 10–5 Clinical photograph shows a tan lesion (malignant lymphoma) with moderate hyperemia elevating the bulbar conjunctiva in an elderly patient.

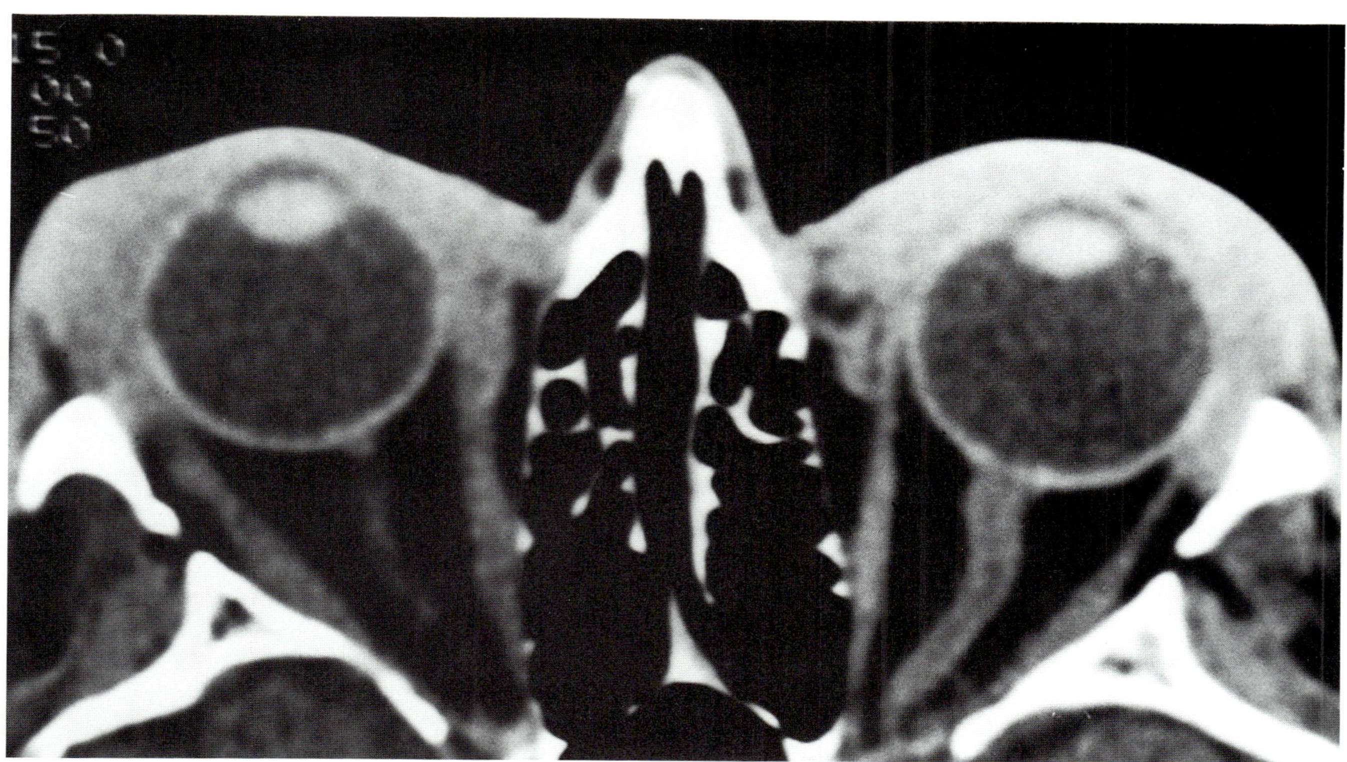

FIGURE 10–6 Photograph of the CT scan shows bilateral lesions that conform anteriorly to the contour of the globe without any orbital bone destruction.

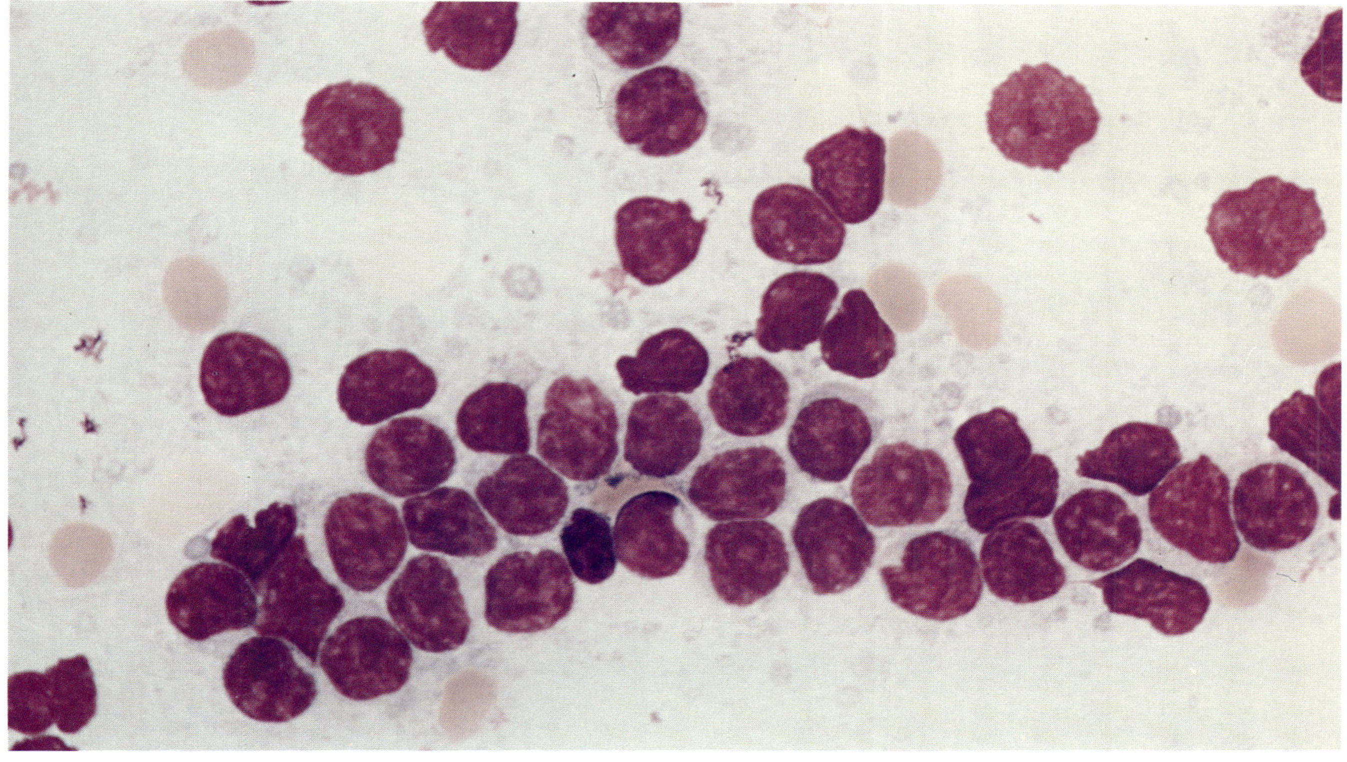

FIGURE 10–7 Photomicrograph shows homogenous population of atypical, but uniform lymphocytic cells with extremely scant cytoplasm. (May-Grünwald Giemsa, × 712)

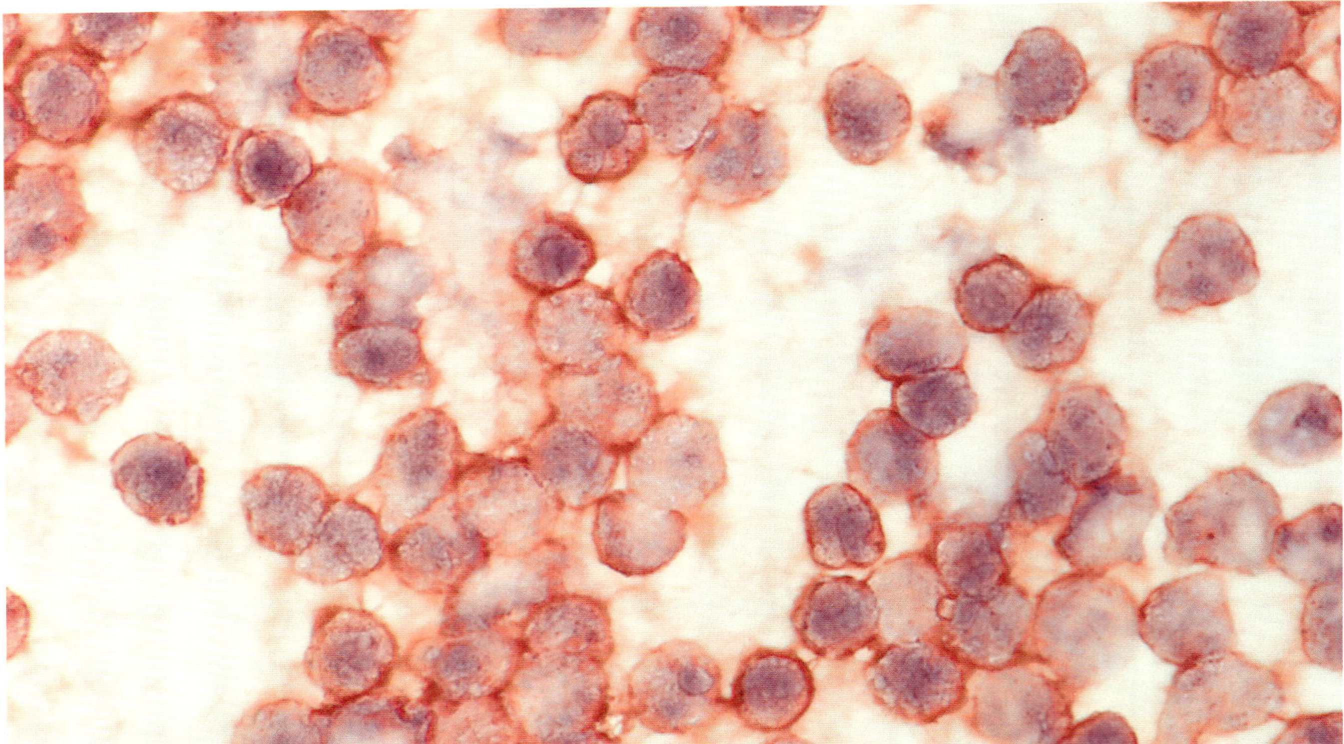

FIGURE 10–8 Photomicrograph shows uniform staining with antibody to κ light chain on cells from an aspiration smear indicative of monoclonality in a case of malignant lymphoma.

Metastatic Small-cell Carcinoma

Small-cell carcinoma is the most common type of lung cancer that metastasizes to the orbit. It is important because orbital findings are sometimes the initial manifestations. The tumor may fill the orbit or be localized to an extraocular muscle (Figure 10–10). The cytologic findings may mimic another small-cell tumor such as lymphoma. Differentiating features include the presence of cell necrosis, nuclear crush artifact, and some variation in cell size. The nuclear crush artifact occurs even in tissue sections and is probably related to increased fragility (Figure 10–11). Electron microscopy of aspiration cytology is useful to demonstrate epithelial elements. Leukocyte markers are helpful to exclude lymphoma.

PRIMARY LACRIMAL TUMORS

The most common primary tumors of the lacrimal gland are pleomorphic adenoma and adenoid cystic carcinoma. Malignant mixed tumor is much less common.

Pleomorphic Adenoma

Pleomorphic adenoma (benign mixed tumor) accounts for about 50% of all epithelial lacrimal gland tumors.[28–30] Benign mixed tumor may occur at any age (mean is 39 years). A painless mass in the lacrimal fossa is often observed. By CT scan, the tumors are globular in shape and may indent sclera and remodel bone. It has been demonstrated that open biopsy of both major salivary and lacrimal gland benign mixed tumors may lead to seeding.[31] No cases of seeding have been documented following fine needle aspiration biopsy of these tumors from salivary glands.[32,33] No cases of fine needle aspiration of lacrimal gland mixed tumor have been reported.[3] Fine needle aspiration of mixed tumors shows tightly clustered benign epithelial cells and a characteristic mucinous chondroid matrix (Figure 10–12).

Adenoid Cystic Carcinoma

Adenoid cystic carcinoma accounts for about 25% to 30% of all epithelial lacrimal gland tumors.[34,35] Clinically, the patients usually have symptoms of pain and develop a mass in the lacrimal fossa (Figure 10–13). On CT scan, adenoid cystic carcinomas are globular tumors, but frequently have irregular destructive margins around orbital bone (Figure 10–14).[36] Fine needle aspiration shows numerous clusters and single cells. The cells form characteristic rosettes that surround magenta basement membrane material (Figure 10–15). This material does not stain with hematoxylin and eosin or Papanicolaou stain. Occasionally, mitotic figures and individual cell necrosis can be seen. This pattern correlates well with basement

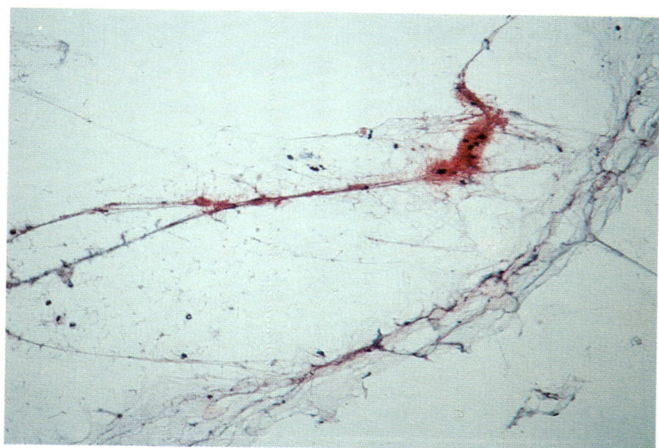

Figure 10–9 Photomicrograph shows scant material with long fibrous tissue strands and scattered lymphocytes. (hematoxylin and eosin, × 48)

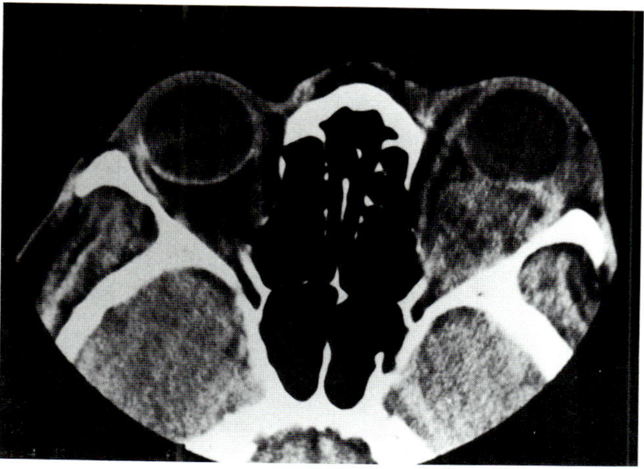

Figure 10–10 Photograph of CT scan shows a tumor that fills the left orbit, pushes the left eye forward (proptosis), and distorts the lateral rectus muscle.

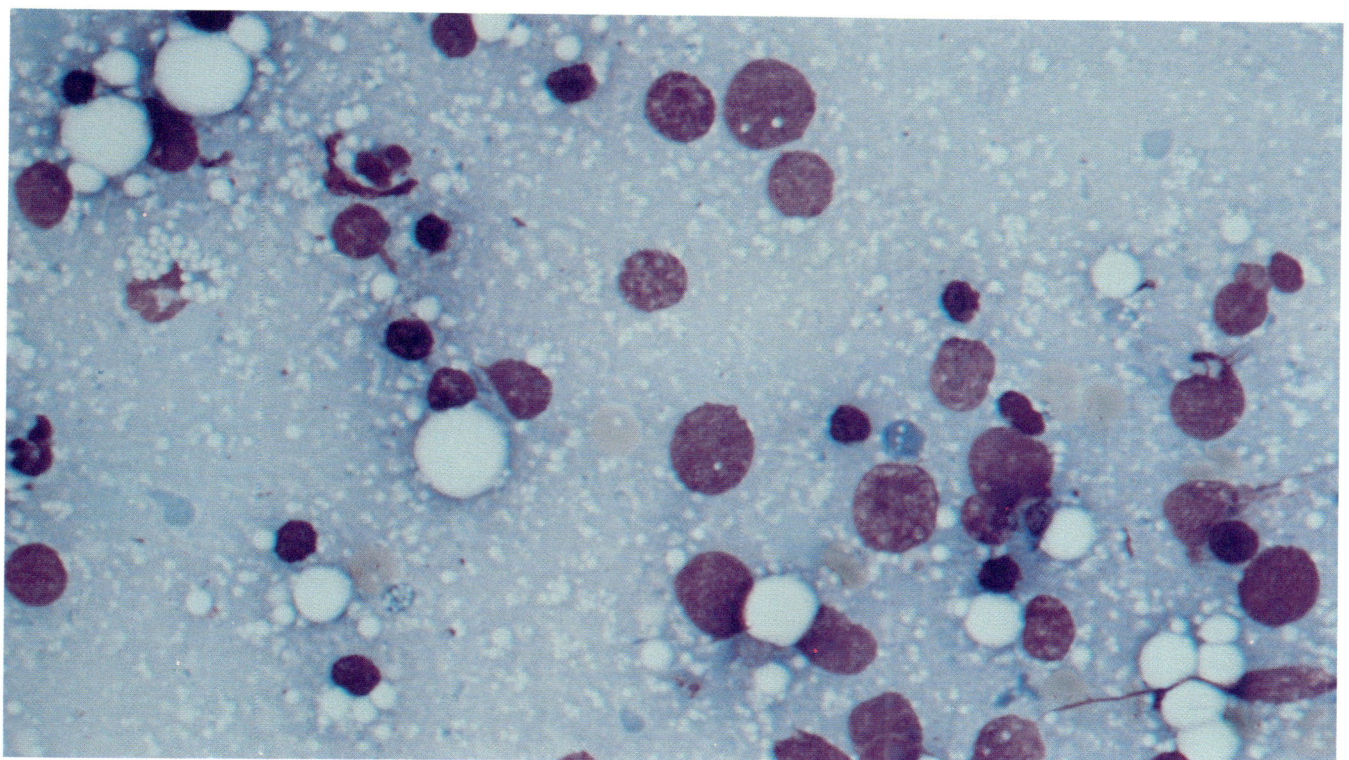

Figure 10–11 Photomicrograph of the fine needle aspirate from the patient in Figure 10–10 shows a small-cell tumor. Some of the cells are necrotic and DNA has been streaked (cells upper left). (May-Grünwald Giemsa, × 712)

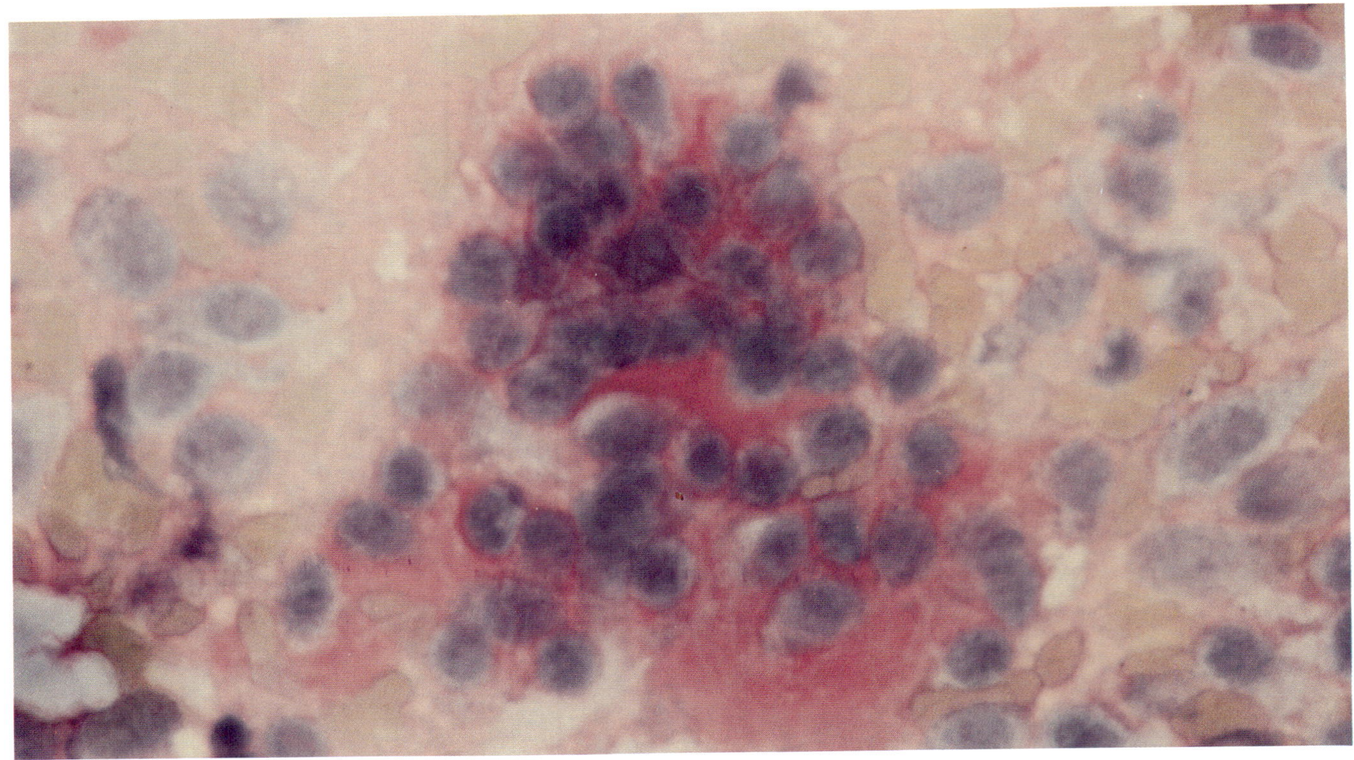

FIGURE 10–12 Photomicrograph shows tightly clustered cells embedded in a red-pink matrix (chondroid) indicative of benign mixed tumor. (May-Grünwald Giemsa, × 125)

FIGURE 10–13 Clinical photograph shows a mass in the right lacrimal fossa in a patient with pain for months.

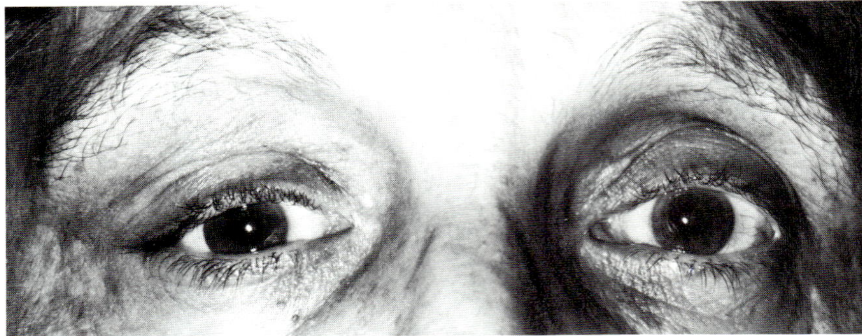

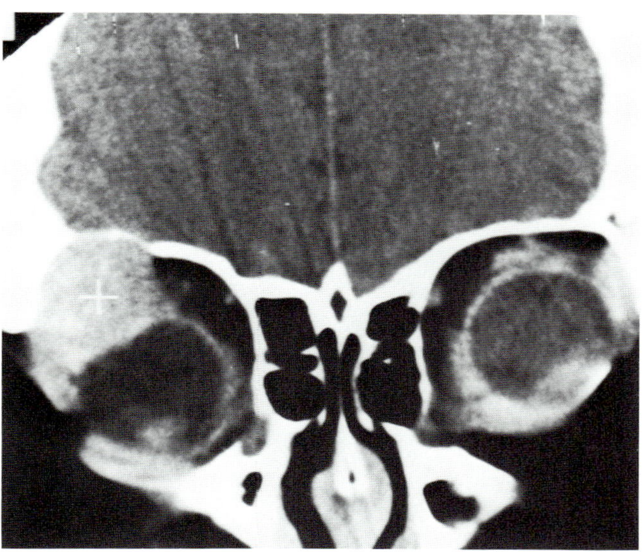

FIGURE 10–14 CT scan from the patient in Figure 10–13 shows destruction of bone at the right superior orbital rim.

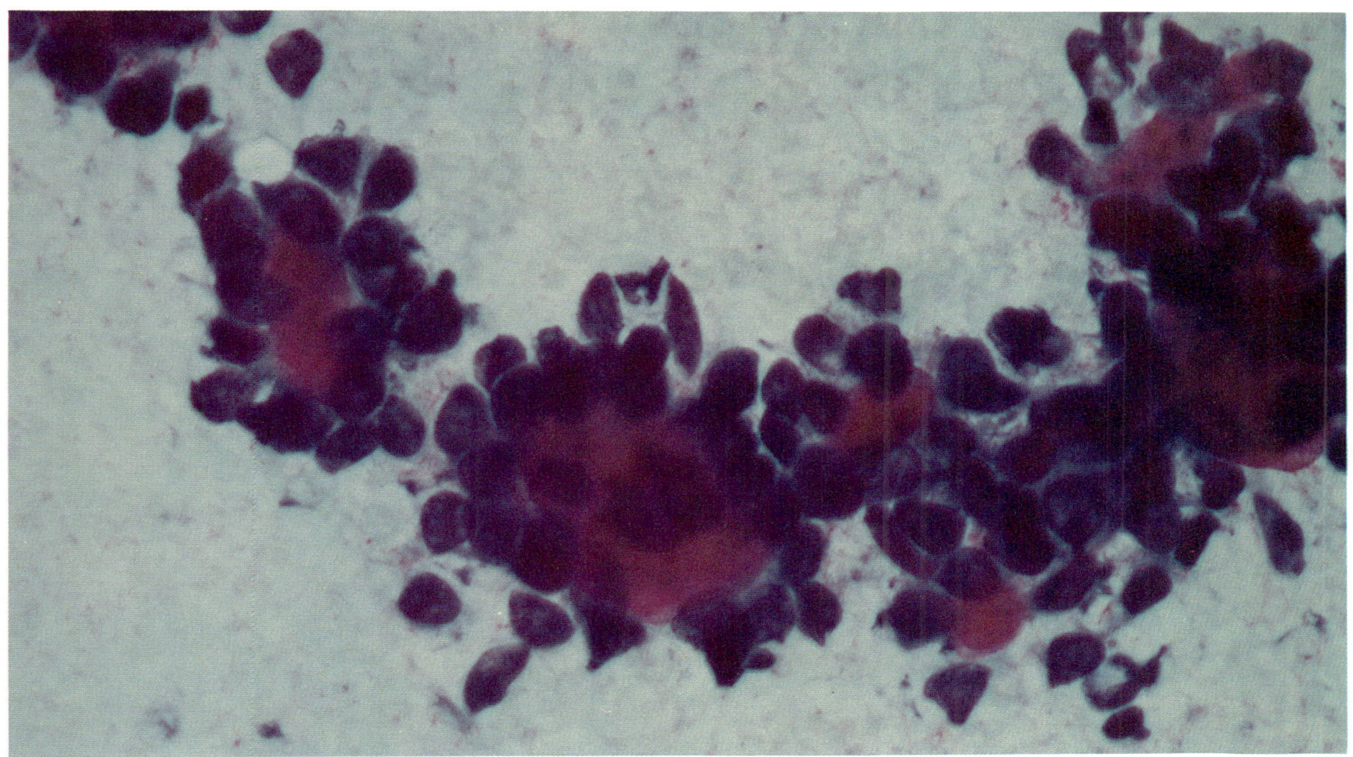

FIGURE 10–15 Photomicrograph shows a fine needle aspiration from the patient shown in Figures 10–13 and 10–14. Tumor cells form rosettes with magenta-purple basement membrane material in the center. The diagnosis was confirmed at the time of orbitectomy. The patient died as a result of postoperative complications. (May-Grünwald Giemsa, ×250)

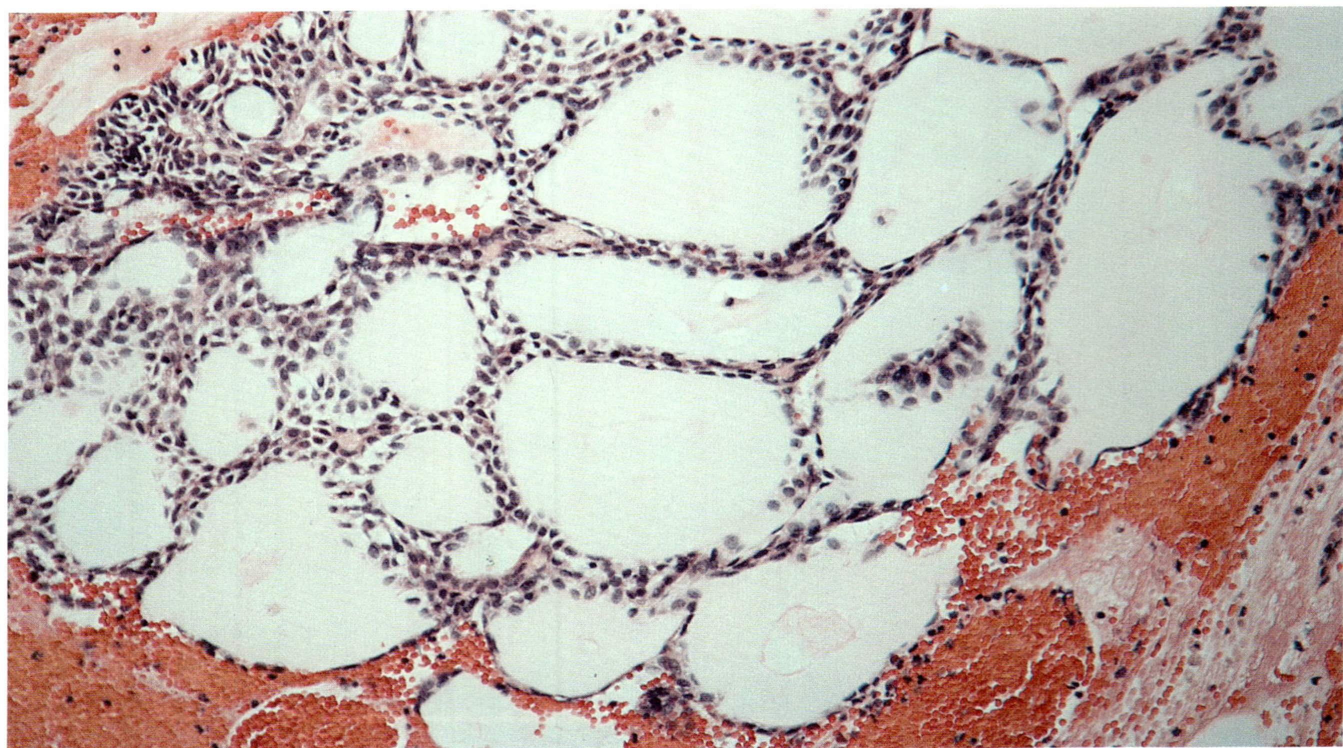

FIGURE 10–16 Photomicrograph shows the classic cribiform pattern of an adenoid cystic carcinoma. (hematoxylin and eosin, × 125)

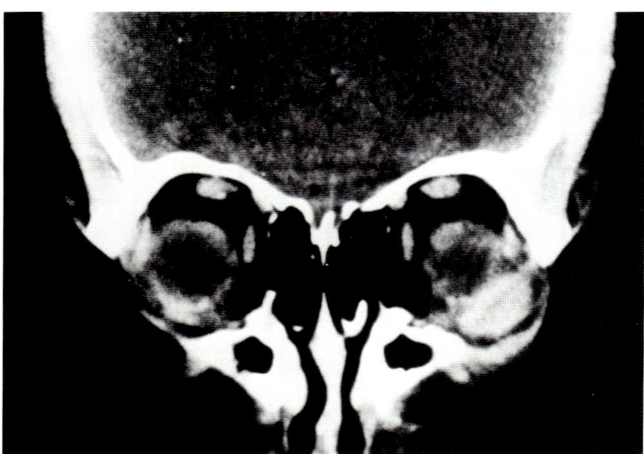

FIGURE 10–17 A 27-year-old male presented with a fleshy mass in the left inferior fornix and a red eye. There was pain and slightly reduced downgaze in the left eye. Photograph of coronal view of the CT scan reveals a globular mass in the infero-temporal orbit.

membrane material seen in the center of cribiform areas on histologic sections (Figure 10–16). The correlation of prognosis of adenoid cystic carcinoma with the histologic patterns is controversial.[37,38] The treatment for these lesions is also controversial and ranges from excision of the tumor with radiation, en bloc excision with resection of contiguous bone, orbital exenteration, and radical orbitectomy.[39,40] At present, there is not enough follow-up data to demonstrate that radical procedures result in cure.

GRANULOMATOUS LESIONS

Some granulomatous lesions of the orbit may be differentiated from each other by fine needle aspiration cytology. Eosinophilic granuloma, ruptured dermoid cyst, and chalazion are so characteristic that the diagnosis can be suggested and usually made definitively with cytology alone. However, in certain cases, ancillary tests are helpful (e.g., electron microscopy for eosinophilic granuloma).

Foreign-body Granuloma

Foreign-body granuloma may present with a subconjunctival or orbital lesion. frequently, patients will not be aware of a previous injury or implanted foreign material. Patients may present with a red eye and reduced eye movements. There may be a yellow infiltrate under the conjunctiva that mimics lymphoma. CT scan may reveal a mass (Figure 10–17).

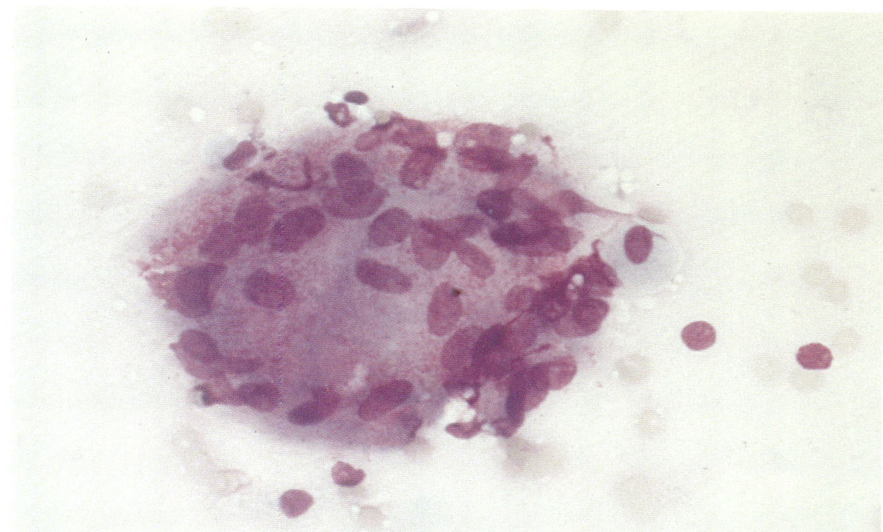

FIGURE 10–18 Photomicrograph of the fine needle (27 gauge) aspirate from the patient in Figure 10–17. Granulomatous inflammation with foreign-body-type giant cells was identified. (May-Grünwald Giemsa, × 250)

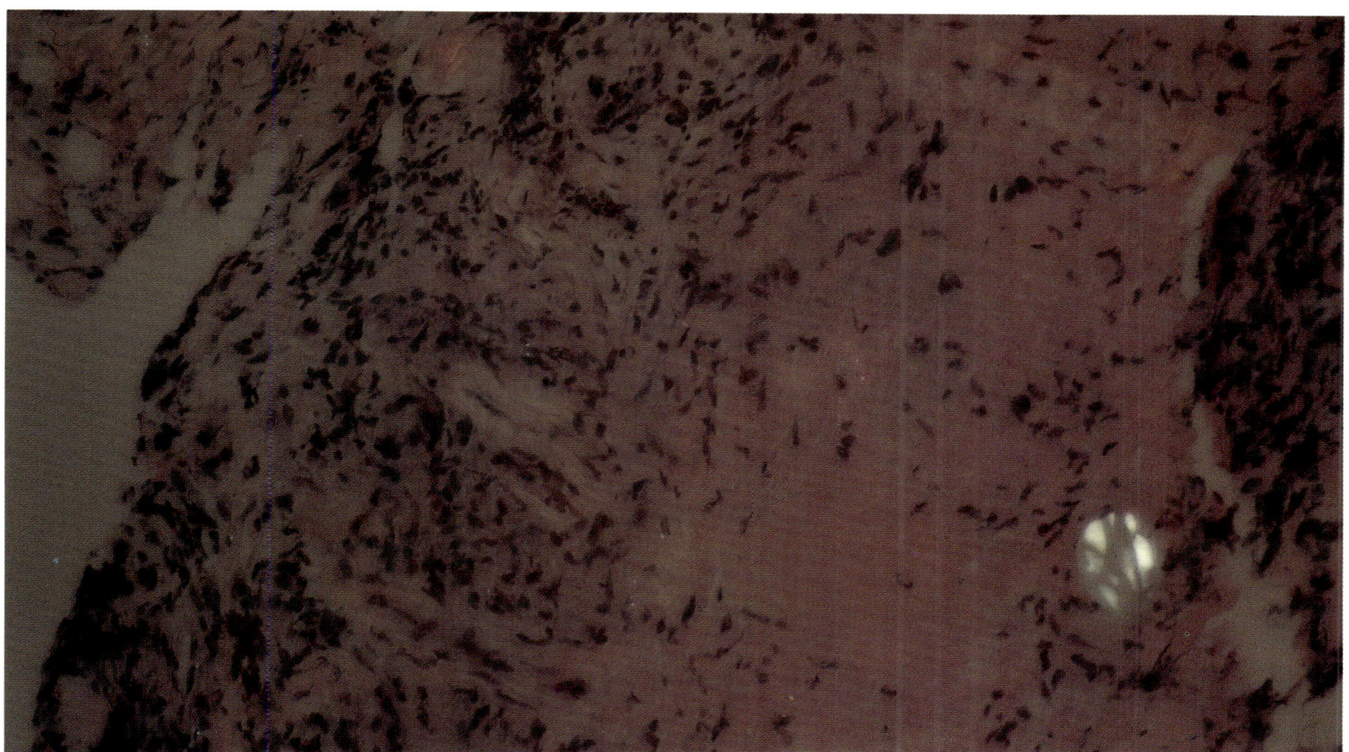

FIGURE 10–19 Photomicrograph shows birefringent foreign material obtained by open biopsy from the patient in Figures 10–17 and 10–18. (hematoxylin and eosin with polarized light, × 43)

Fine needle aspiration demonstrates a granuloma with foreign-body-type giant cells (Figure 10–18). Infectious causes of granulomatous inflammation should be excluded with culture and special stains. This can be accomplished by fine needle aspiration.[41] Biopsy is usually necessary to uncover the foreign body (Figure 10–19).

Chalazion

Chalazia are lipogranulomatous reactions that occur in the eyelid because of meibomian gland obstruction. They may be associated with infection or neoplasms, but

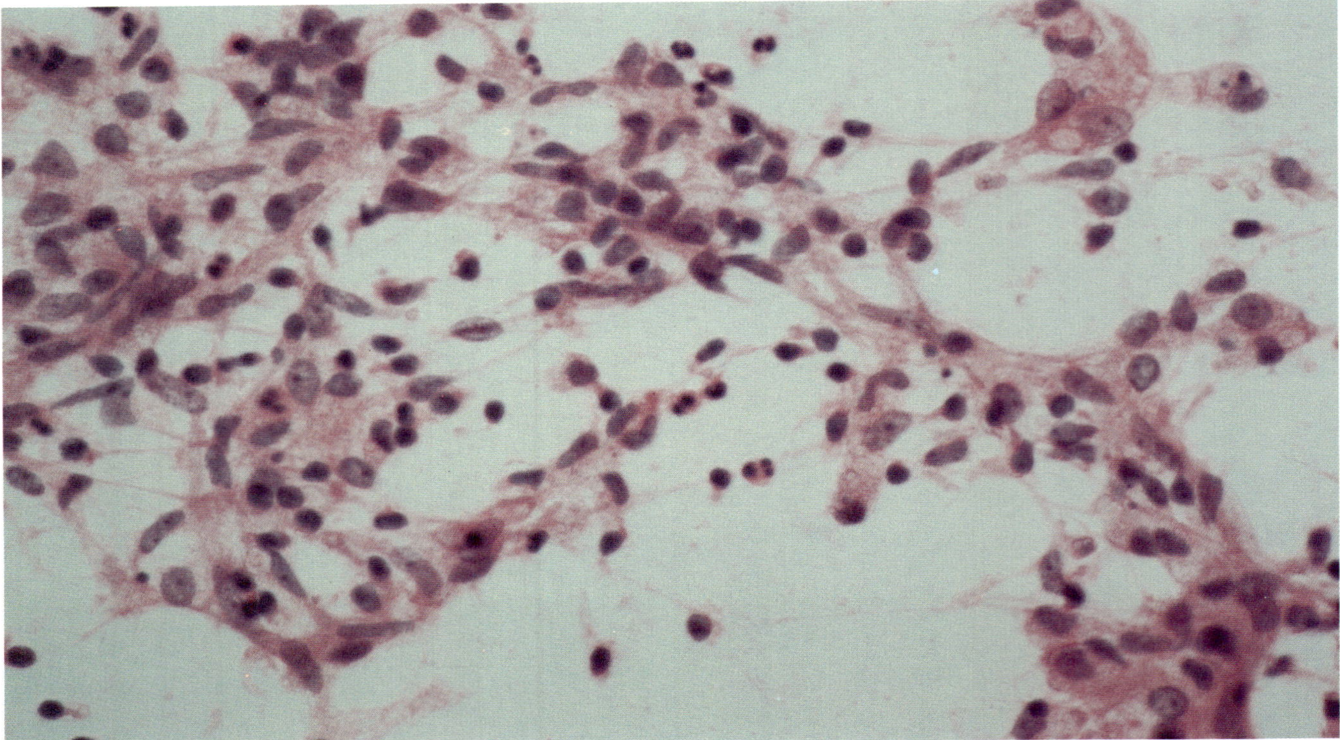

FIGURE 10–20 Photomicrograph shows numerous histiocytes containing lipid vacuole. The histiocytes are uniform in character with round nuclei. (hematoxylin and eosin, × 125) Diagn Cytopathol 1991;7:132–143. Reprinted courtesy of Wiley Liss.

are usually secondary to inspissated secretions.[42] Usually, the diagnosis is obvious clinically and no biopsy is necessary. In longstanding cases, the chalazion may present as a discrete mass. Fine needle biopsy may be done to rule out sebaceous carcinoma or abscess.[3,43] Smears of the aspirate show numerous histiocytes with foamy cytoplasm and occasional granulation tissue (Figure 10–20). Treatment of this lesion may include observation or removal by curettage.

Ruptured Dermoid Cyst

Dermoid cysts are congenital lesions that represent arrested migration of ectoderm entrapped in orbital soft tissue or between sutures of orbital bone.[44] They are the most common orbital tumor in children. It is preferable not to aspirate a dermoid cyst to avoid spillage of its contents.[3] However, deep dermoids are frequently difficult to diagnose clinically. This type of dermoid cyst presents in adulthood as a slowly growing mass in the supero-temporal orbit. There is often extension through bone sutures that can be confused with boney erosion of lacrimal gland tumors.[45] Seven fine needle aspiration biopsies of dermoid cysts have been reported and all presented as adults and at least two of them had bone involvement.[3,46] In at least two of the cases, there was a history of a tumor since childhood and the diagnosis was easily made by fine needle aspiration.[46,47] In others, malignancy was clinically suspected. In one case, the cytologic findings were even reported as malignant (a false positive), but details are not disclosed.[9] Fine needle aspiration of an unruptured dermoid cyst contains anucleate squamous cells, keratin debris, and, occasionally, hair shafts. A ruptured dermoid cyst will also contain granulomatous inflammation with multinucleated giant cells (Figures 10–21 and 10–22). We have not seen adnexal structures of the cyst wall in the four dermoid cysts we have examined cytologically.

Wegener's Granulomatosis

Wegener's granulomatosis is characterized by necrotizing vasculitis and granulomatous inflammation in the upper respiratory tract, lung, and kidneys. It occurs predominantly in males. Orbital involvement occurs in about 20% of the cases and is usually bilateral.[48] It is extremely difficult to make a specific diagnosis of Wegener's granulomatosis or even a diagnosis of necrotizing vasculitis by fine needle aspiration. Usually, extremely scant material is obtained, but histiocytes and groups of necrotic cells can sometimes be identified (Figure 10–23). These lesions frequently have extensive fibrosis, and collection of adequate material by fine needle biopsy is very difficult (Figure 10–24).

Sarcoidosis

Sarcoidosis is a multisystem, granulomatous disease of unknown cause. It occurs more frequently in women and blacks. The lacrimal gland may be palpably enlarged in 7% of patients. Patients may be considered to have a lacrimal gland tumor (Figure 10–25). Fine needle biopsy demonstrates granulomatous inflammation without necrosis. Multinucleated giant cells are usually evident (Figure 10–26). When a fine needle aspirate reveals granulomatous inflammation on the first biopsy, another aspirate should be considered to obtain culture and special stains to rule out infectious causes. Sarcoidosis is a clinical diagnosis and cannot be made by cytologic or histologic findings alone.

Eosinophilic Granuloma

Eosinophilic granuloma is one of the spectrum of the diseases known as histiocytosis X, which includes Hand-Schüller-Christian disease, Letterer-Siwe disease, and eosinophilic granuloma. It is thought that proliferating Langerhans' cells are responsible for the lesion. The orbit is most often involved by unifocal disease, eosinophilic granuloma.[49,50] The disease usually occurs in children and teenagers and involves bone and adjacent soft

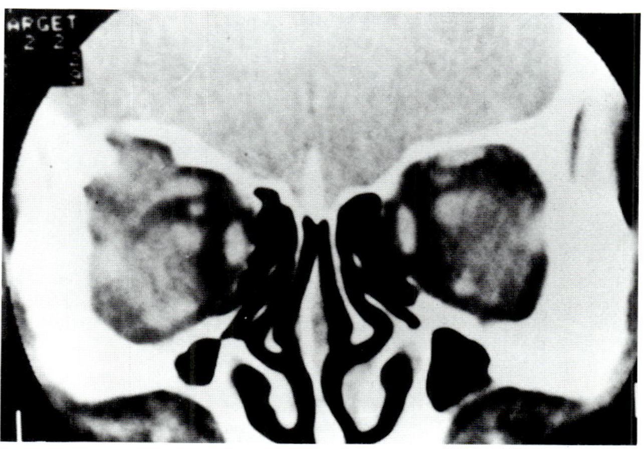

FIGURE 10–21 A 30-year-old female presented with pain and headaches. Photograph of the CT scan shows a tumor in the orbit that is eroding the orbital roof. Adenoid cystic carcinoma was suspected clinically.

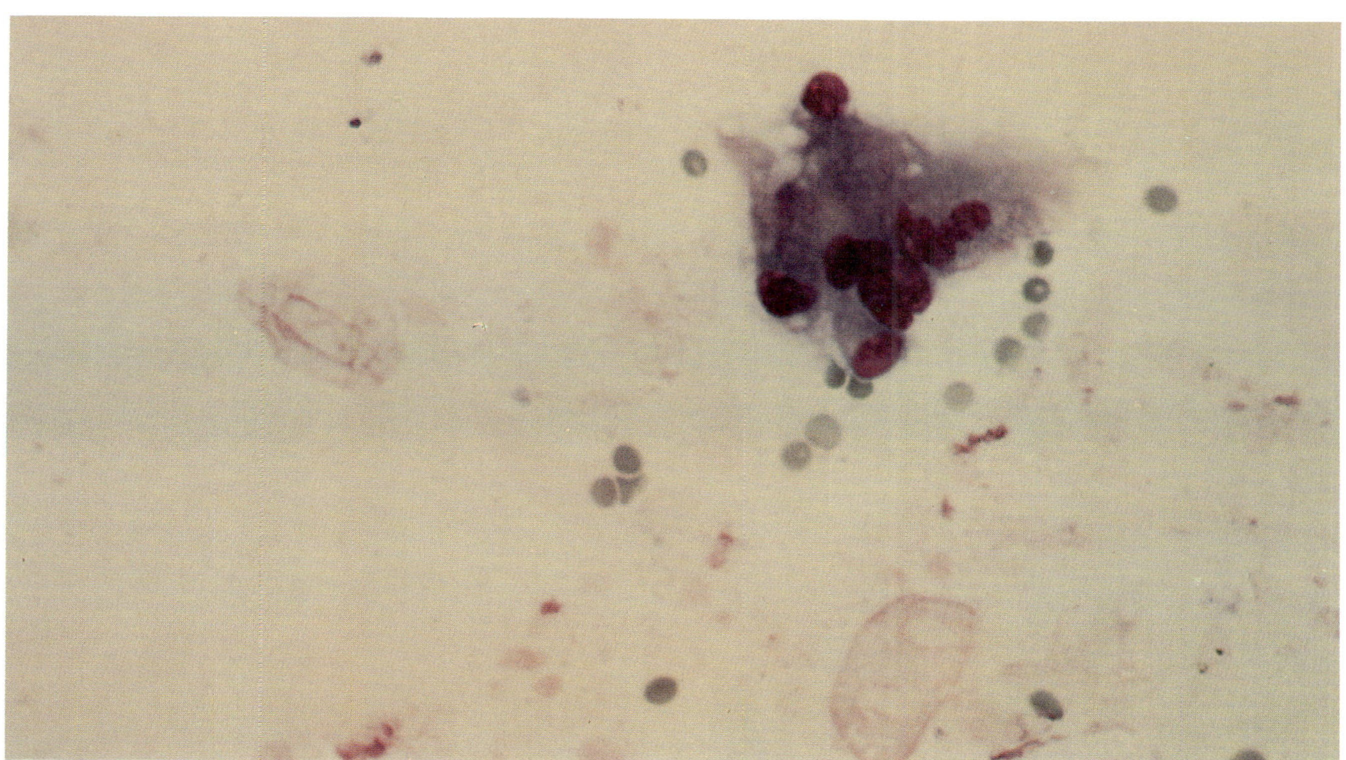

FIGURE 10–22 Fine needle aspiration from the patient shown in Figure 10–21 shows an anucleate squamous cell and multinucleated giant cell consistent with a ruptured dermoid cyst. (May-Grünwald Giemsa, × 200)

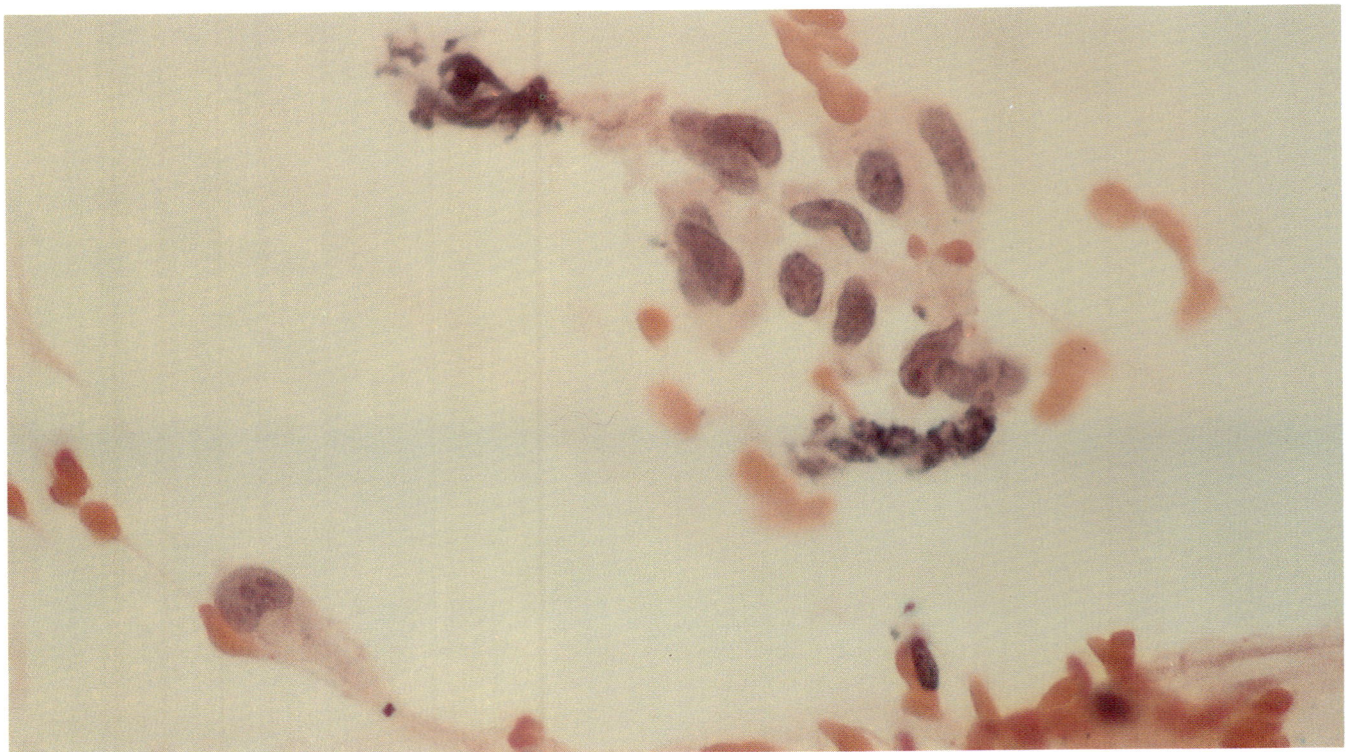

FIGURE 10–23 (ABOVE) A 62-year-old female presented with a history of pulmonary lesions and a previous biopsy diagnosis of necrotizing vasculitis. Photomicrograph shows a collection of histiocytes adjacent to a cluster of necrotic cells. These findings are indicative of necrotizing granulomatous inflammation, but are not diagnostic for Wegener's granulomatosis. An open biopsy is necessary to demonstrate the vasculitis. (hematoxylin and eosin, × 400)

FIGURE 10–24 (BELOW) Photomicrograph from section of the biopsy specimens from the patient in Figure 10–23 shows a necrotizing vasculitis with granulomatous inflammation and rare eosinophils. (hematoxylin and eosin, × 400)

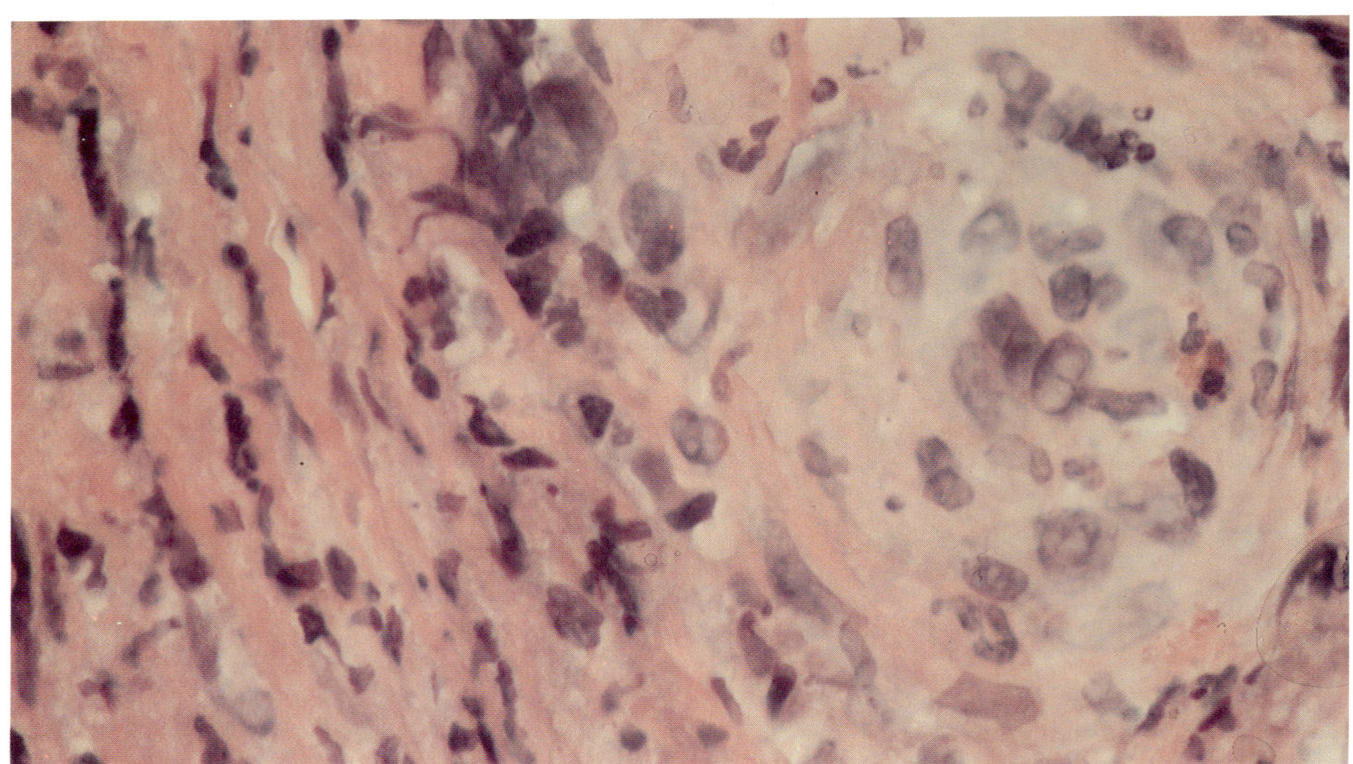

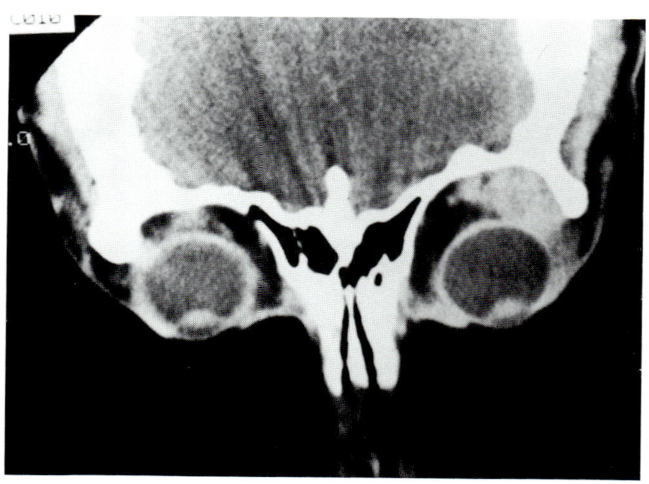

FIGURE 10–25 A 52-year-old black female presented with pain and irritation in the left eye. Photograph of the CT scan shows a mass in the lacrimal fossa, clinically suspected of being lymphoma.

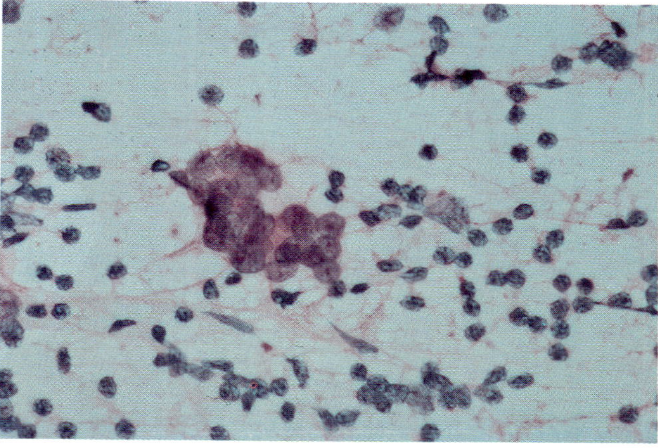

FIGURE 10–26 Photomicrograph of fine needle aspiration shows a moderately dense, but heterogenous population of lymphocytes and multinucleated giant cells. (hematoxylin and eosin, × 400)

tissue (Figures 10–27 and 10–28). It is common in the superotemporal portion of the orbit. The diagnosis can be readily made by fine needle aspiration provided sampling is adequate.[51] Langerhans' cells are evident as histiocytes with grooved or indented nuclei. Multinucleated giant cells, eosinophils, and neutrophils are present (Figure 10–29). The diagnosis can be confirmed by electron microscopy of the aspiration specimen (Figure 10–30). Treatment usually includes curettage.

EPITHELIAL LESIONS

Squamous Carcinoma

Squamous carcinoma is the most common paranasal sinus tumor to invade the orbit.[52] The maxillary sinus is the original site of the carcinoma in the majority of cases. The patients with this tumor may present predominantly with orbital signs. In these cases, destruction of the orbital floor is usually seen on CT scans (Figure 10–31).

Fine needle aspiration reveals abundant malignant cells. The key to the diagnosis is the discovery of squamous differentiation. Frequently, the tumors will have a spindle-cell appearance (Figure 10–32). Some of these tumors may arise from inverted papillomas (Figure 10–33). Patients with squamous carcinoma from a sinus involving the orbit in general, have a poor prognosis.

Sebaceous Carcinoma

Sebaceous carcinoma originates from meibomian glands and glands of Zeis in the eyelid. It may present clinically in different forms, a small yellow nodule, a diffuse thickening of the eyelid, or a mass in the lacrimal fossa.[53] As a small yellow nodule, it is frequently misdiagnosed as a chalazion. As a diffuse thickening of the eyelid, it may be misdiagnosed as blepharitis (Figure 10–34). As an orbital mass, it may be misdiagnosed as a primary lacrimal gland tumor (Figure 10–35).[54]

Fine needle aspiration of sebaceous carcinoma has been reported in numerous cases for eyelid tumors.[43,55] Fine needle aspiration is generally done when an orbital mass is the predominant presenting feature or if the abnormalities of conjunctiva and eyelid are overlooked. Fine needle aspiration shows abundant material with large cells and numerous lipid vacuoles (Figure 10–36). As a result of fine needle aspiration, the surgeon may plan to do multiple eyelid and conjunctiva biopsies to determine the extent of the tumor because independent foci of sebaceous carcinoma in the eyelid have been noted in up to 10% of cases (Figure 10–37).[56,57]

Basal-cell Carcinoma

Basal-cell carcinoma is the most common malignant epithelial tumor of the eyelid.[58] Fine needle aspiration is unnecessary to diagnose most primary lesions because skin biopsy is so easily performed. Occasionally, recurrent deep orbital lesions present as orbital masses. Fine needle aspiration of basal-cell carcinoma shows tight clusters of small epithelial cells with atypical nuclei and occasional palisading (Figure 10–38).There is a high rate of negative and insufficient biopsies with basal cell carcinoma.

Metastatic Carcinomas

A variety of metastatic carcinomas initially present with orbital manifestations.[59] Metastatic breast, renal cell, transitional cell, and prostate carcinomas have all been specifically identified by orbital fine needle aspiration, but most are only identified as adenocarcinoma. Immunocytochemical studies may be helpful in specifying some sources of origin, such as prostate.[60]

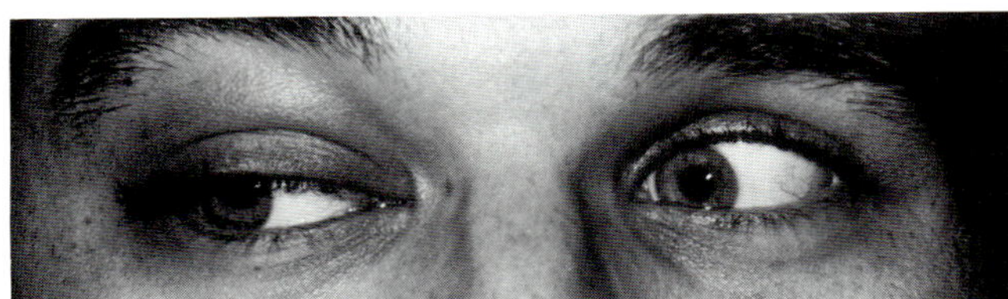

FIGURE 10–27 Clinical photograph of a 23-year-old male who presented with a six-week history of headache and decreased vision. There is right orbital soft-tissue swelling and restriction of lateral gaze of the right eye.

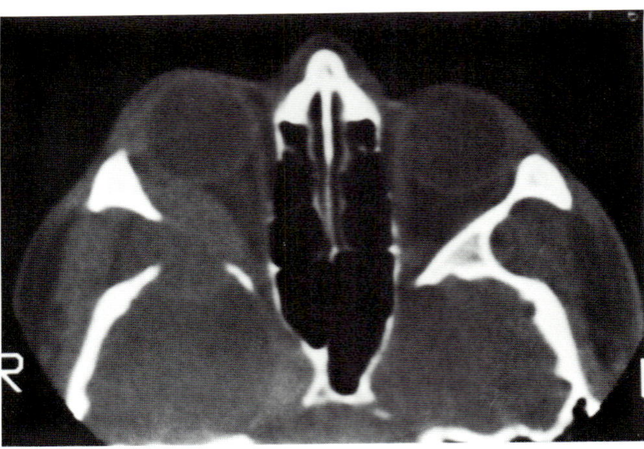

FIGURE 10–28 Photograph of the CT scan shows a mass that is centered in and has destroyed the greater wing of the sphenoid. Reprinted with permission of the publisher from Glasgow BJ and Layfield LJ. Fine needle aspiration biopsy of orbital and periorbital masses. Diagn Cytopathol 1991;7:132–143.

FIGURE 10–29 Photomicrograph from the patient in Figure 10–28 shows multinucleated giant cells, eosinophils, and lobated histiocytes with nuclear grooves characteristic of eosinophilic granuloma. (Papanicolaou, × 250)

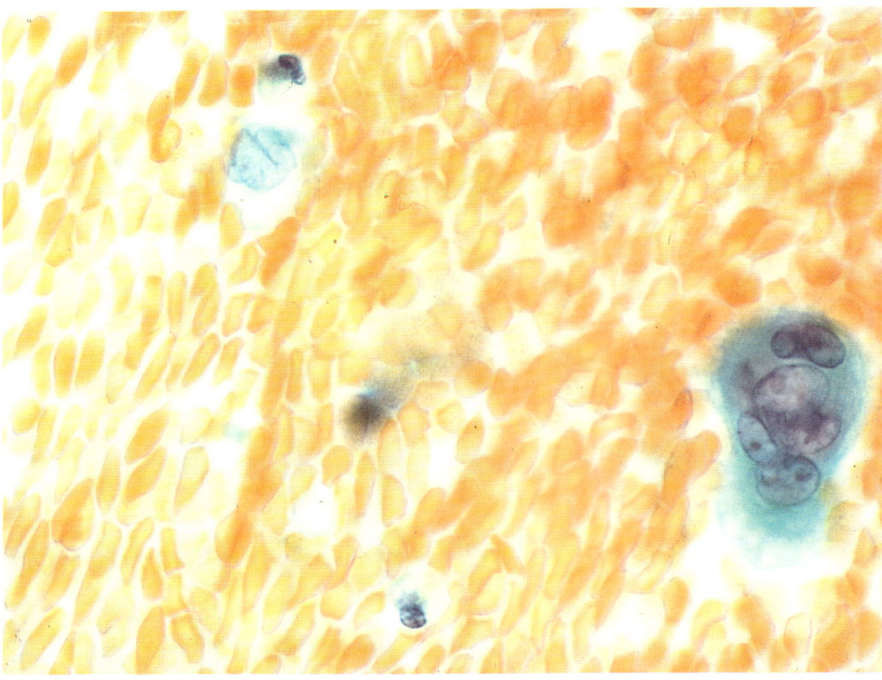

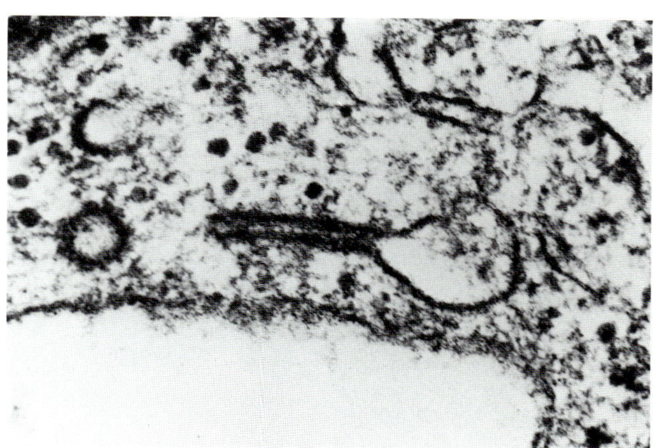

FIGURE 10–30 Electron micrograph from the specimen from the patient in Figure 10–27 shows the racquet-shaped bilamellar structure, also known as the Birbeck granule that is characteristic of the histiocytosis X syndrome. Reprinted with permission of the publisher from Glasgow BJ and Layfield LJ. Fine needle aspiration biopsy of orbital and periorbital masses. Diagn Cytopathol 1991;7:132–143.

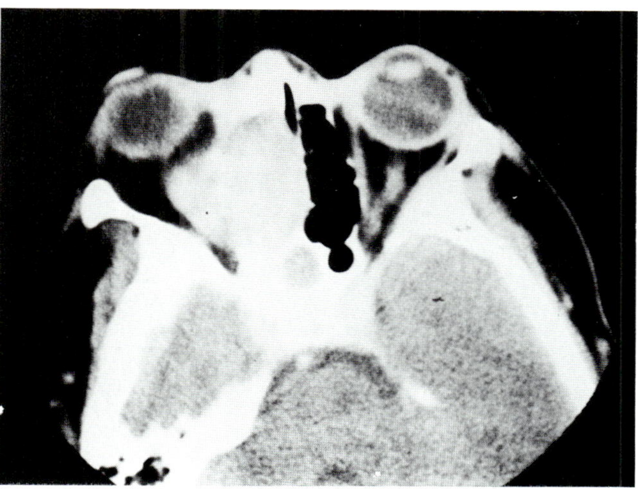

FIGURE 10–31 Photograph of the CT scan shows a mass that fills the orbit and has caused bone destruction in the ethmoid sinus and orbital floor. The radiologic differential diagnosis at the time of fine needle aspiration included mucormycosis as a leading choice.

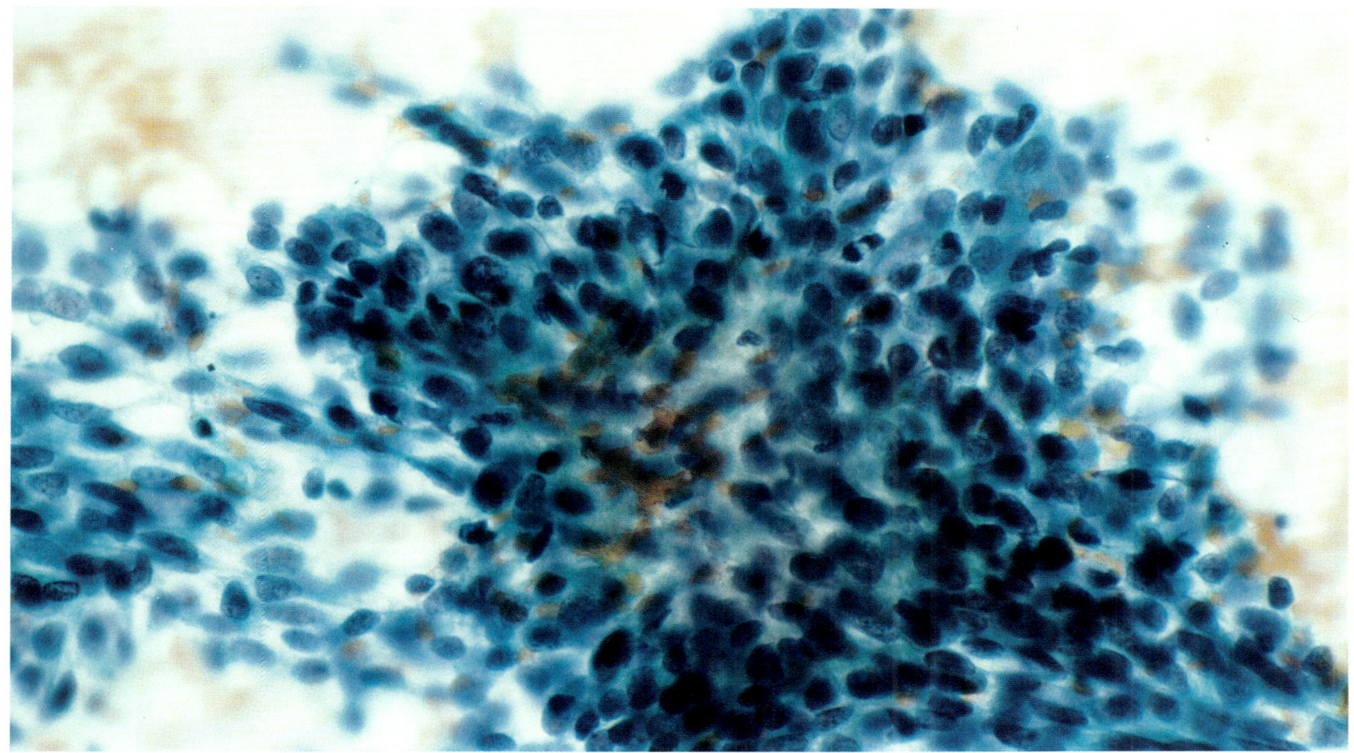

FIGURE 10–32 Clusters of papillary cells were observed and were somewhat spindled in places; numerous foci of necrosis and many mitotic figures were observed. Focal keratinization, seen as orangeophilia on Papanicolaou stain, was identified. (hematoxylin and eosin, × 250)

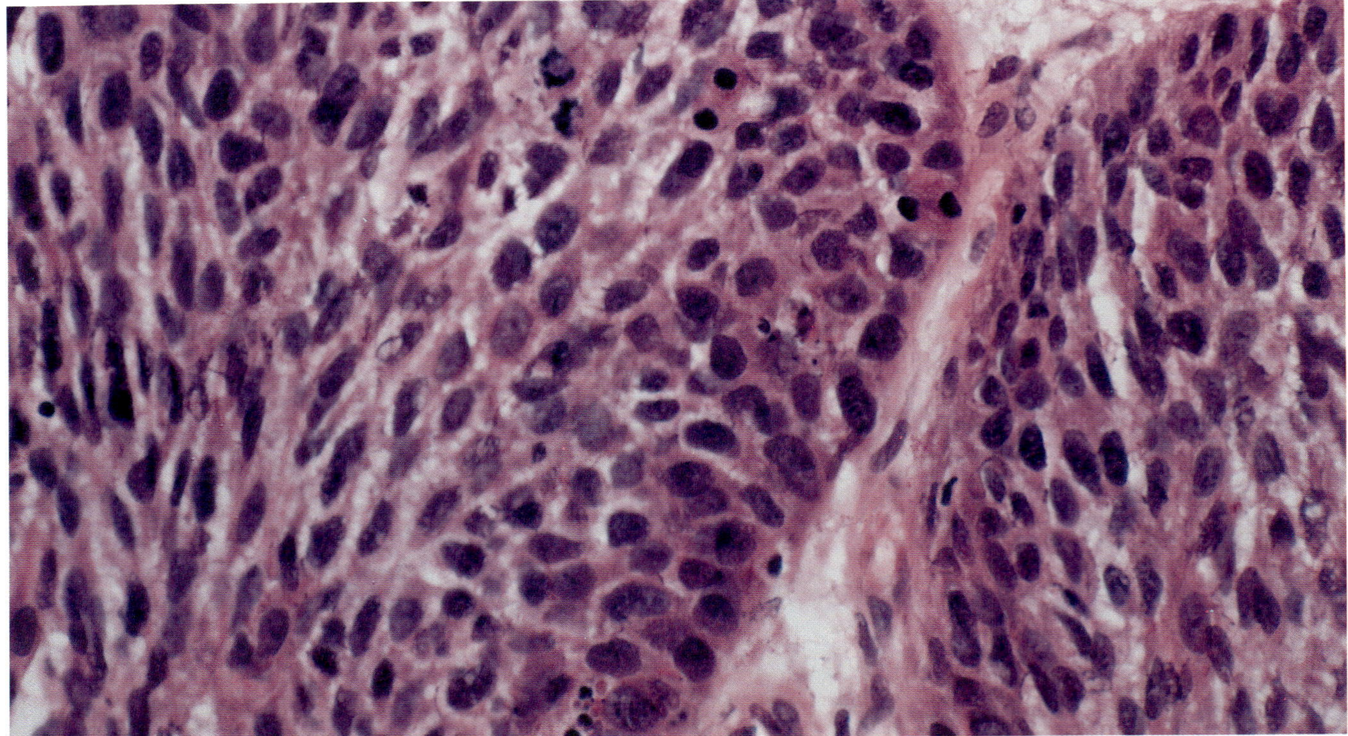

Figure 10–33 Photomicrograph of a histologic section from the patient shown in Figure 10–12 shows a papillary architecture with marked atypia indicative of a squamous carcinoma. (hematoxylin and eosin, × 250)

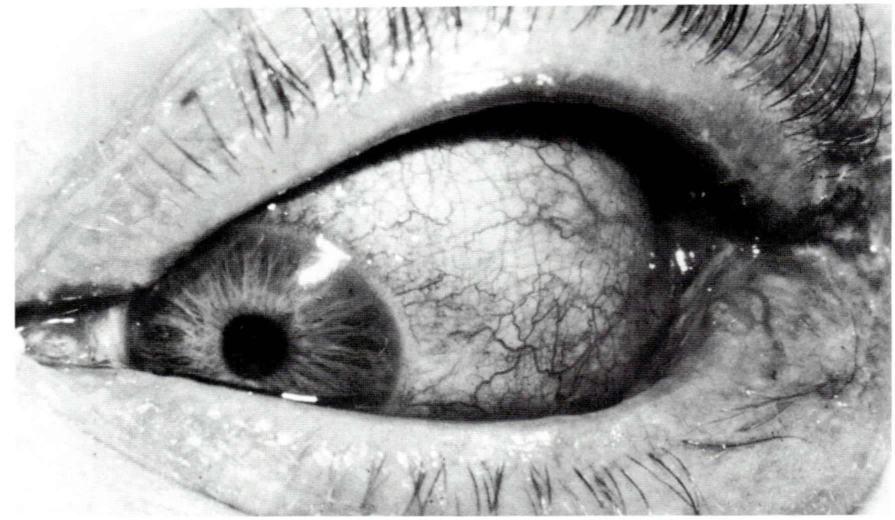

Figure 10–34 A 62-year-old male presented with a history of a chalazion that did not respond to conservative therapy after months of treatment. Diffuse thickening of the lower eyelid is visible in this clinical photograph. Reprinted with permission of the publisher from Glasgow BJ and Layfield LJ. Fine needle aspiration biopsy of orbital and periorbital masses. Diagn Cytopathol 1991;7:132–143.

PERINEURAL LESIONS

Schwannoma (Neurilemmoma)

Schwannomas are tumors of peripheral nerves that presumably arise from Schwann cells. They occur between the ages of 20 and 50 years and are associated with neurofibromatosis.[61] Clinically, patients frequently exhibit proptosis, lid swelling, diplopia, and indentation of the posterior sclera (Figure 10–39).[62] CT scan reveals that neurilemmomas have round smooth borders and expand cortical bone (Figure 10–40). Fine needle aspiration of schwannoma is usually not diagnostic because scant material is obtained. However, myxoid areas associated with spindle cells and bent nuclei are suggestive of a neural lesion (Figure 10–41). Schwannomas are encapsulated and have a firm consistency (Figure 10–42). The tightly packed spindle cells form characteristic Antoni A and B areas that are difficult to discern in fine needle aspiration specimens (Figure 10–43). The cohesive architecture of the schwannoma accounts for the inadequate specimens previously reported.[8]

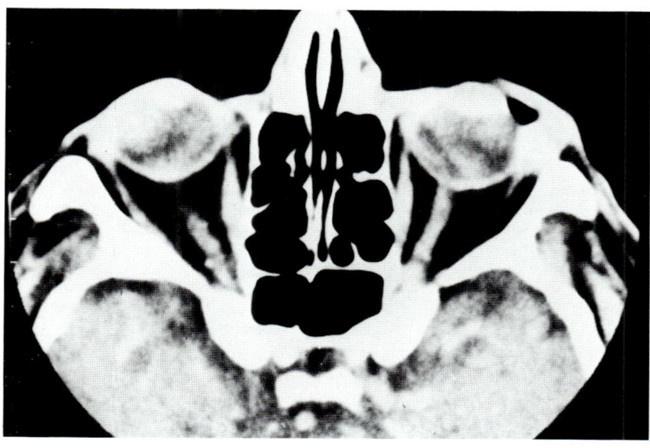

FIGURE 10–35 CT scan of the orbit from the patient in Figure 10–34 shows an orbital mass that is contiguous with the eyelid. Reprinted with permission of the publisher from Glasgow BJ and Layfield LJ. Fine needle aspiration biopsy of orbital and periorbital masses. Diagn Cytopathol 1991; 7:132–143.

Meningioma

Meningiomas in the orbit presumably arise from the arachnoid tissue of the optic nerve and the meninges adjacent to the sphenoid and nearby intracranial struc-

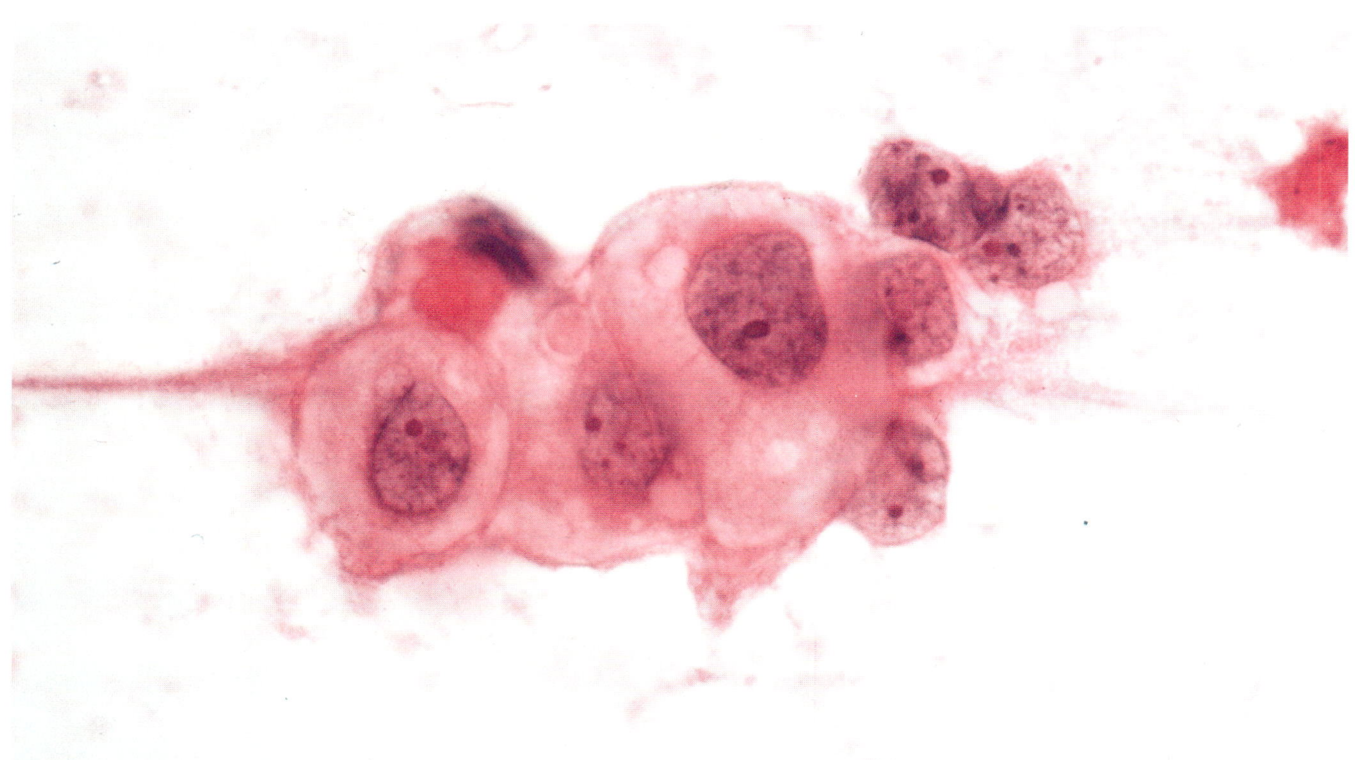

FIGURE 10–36 Fine needle aspiration from the patient in Figures 10–34 and 10–35 demonstrates a cluster of enlarged cells with abundant vacuolated cytoplasm, pleomorphic nuclei with prominent nucleoli indicative of sebaceous carcinoma. Reprinted with permission of the publisher from Glasgow BJ and Layfield LJ. Fine needle aspiration biopsy of orbital and periorbital masses. Diagn Cytopathol 1991;7:132–143.

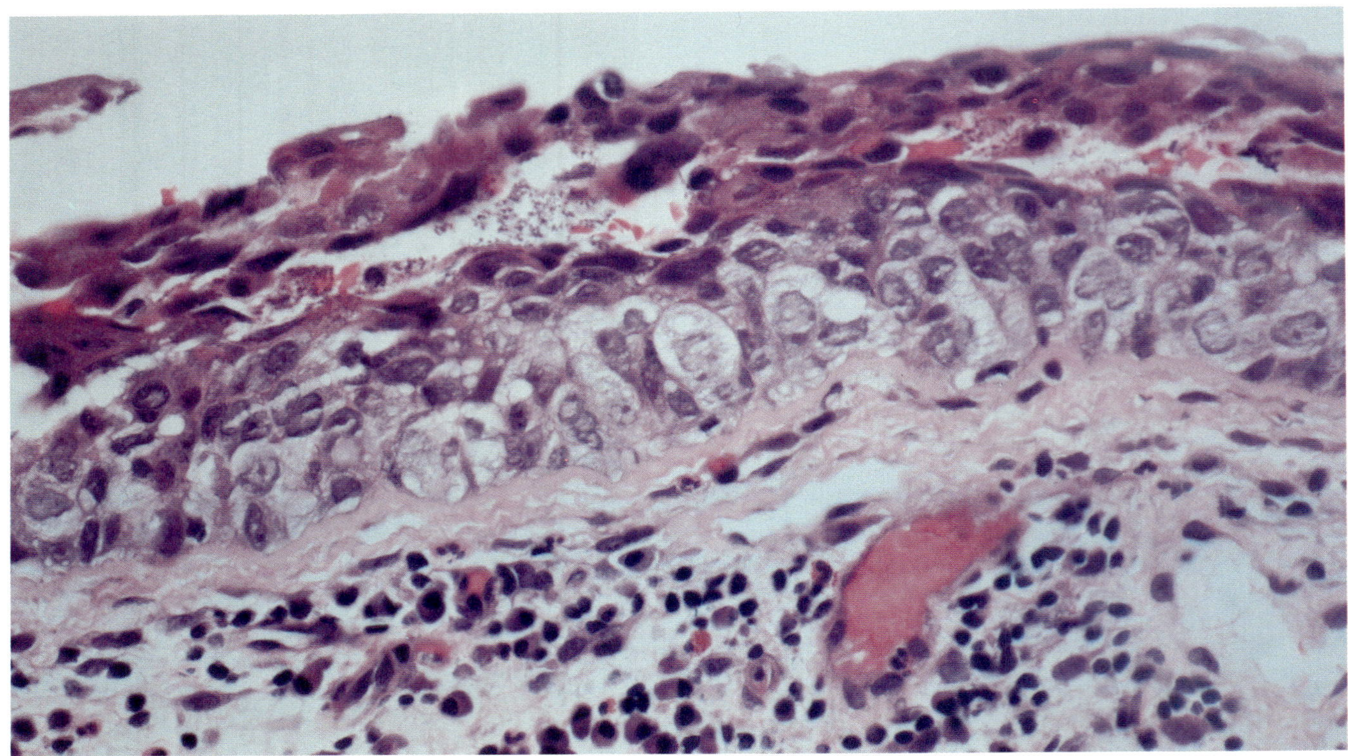

FIGURE 10–37 (ABOVE) Conjunctival biopsy from the patient shown in Figures 10–34, 10–35, and 10–36 shows nests of tumor cells at the base and within the epithelium. (hematoxylin and eosin, × 125)

FIGURE 10–38 (BELOW) Photomicrograph shows a suggestion of peripheral palisading around tight clusters of epithelial cells. Individual cells have round nuclei and small nucleoli. Courtesy of Associate Professor Lester J. Layfield, M.D., University of Iowa.

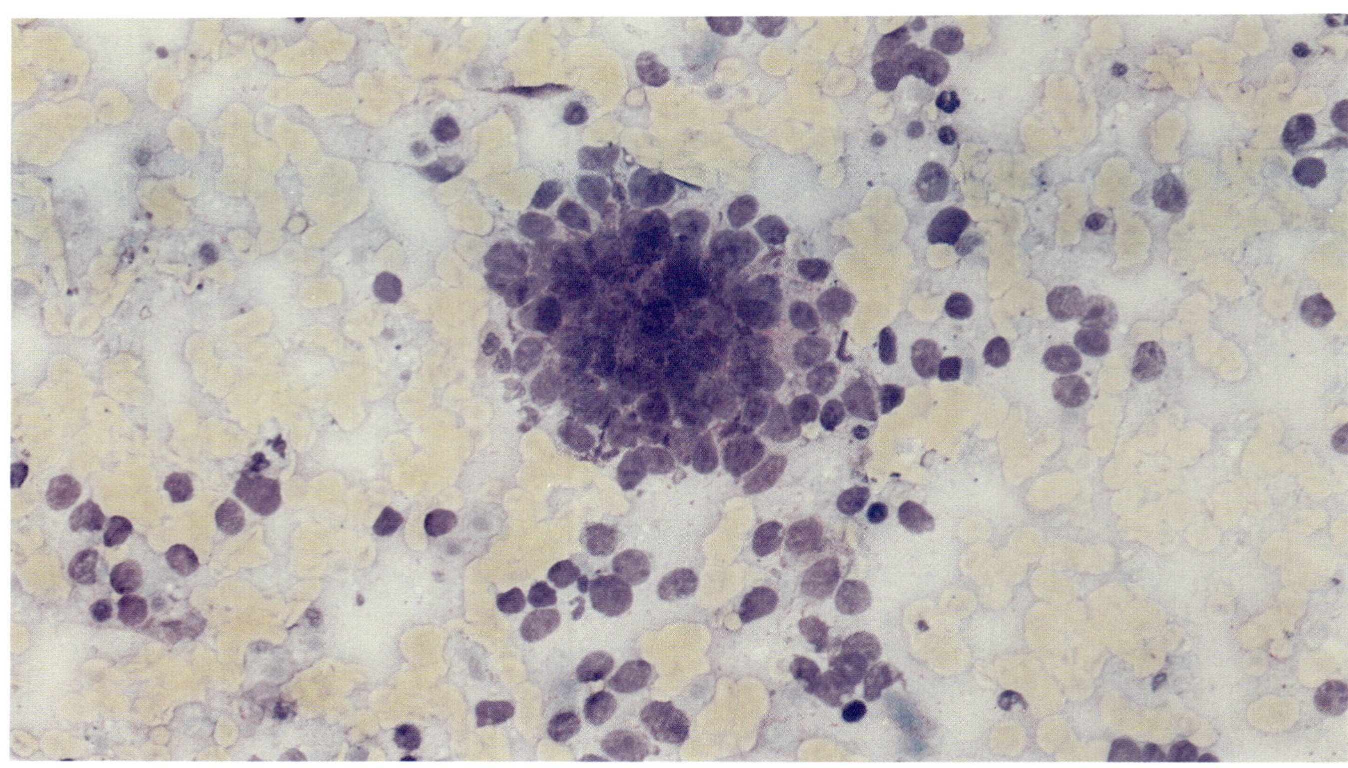

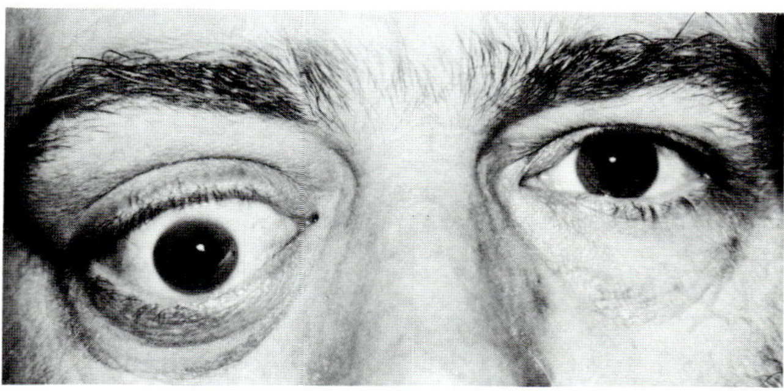

FIGURE 10–39 A 39-year-old male with a six-month history of proptosis and marked inferior and nasal displacement of the right eye caused by a schwannoma. Despite the longstanding process, the eye remains relatively uninflamed. Reprinted with permission of the publisher from Glasgow BJ and Layfield LJ. Fine needle aspiration biopsy of orbital and periorbital masses. Diagn Cytopathol 1991;7:132–143.

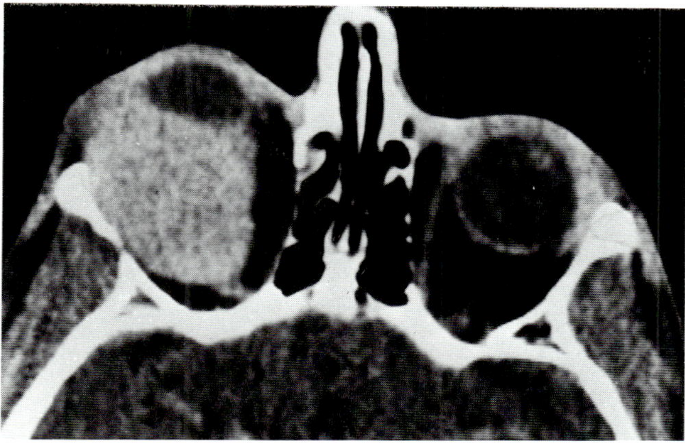

FIGURE 10–40 CT scan from the patient in Figure 10–39 shows an intraconal mass that expands the greater wing of the sphenoid and pushes the eye forward. Reprinted with permission of the publisher from Glasgow BJ and Layfield LJ. Fine needle aspiration biopsy of orbital and periorbital masses. Diagn Cytopathol 1991;7:132–143.

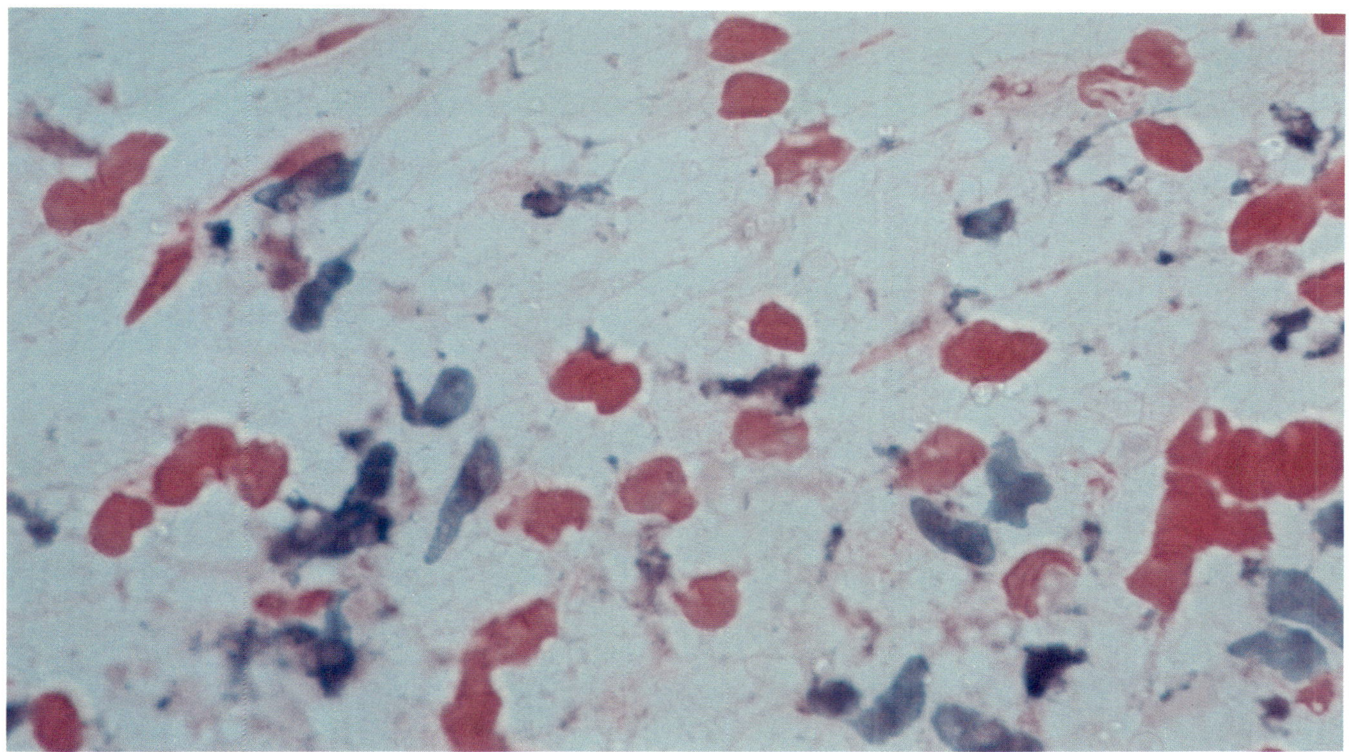

FIGURE 10–41 Fine needle aspiration of the patient in Figures 10–39 and 10–40 shows numerous spindle cells, a myxoid background, and bent nuclei. These findings are suggestive of a neural tumor. (hematoxylin and eosin, $\times$ 400)

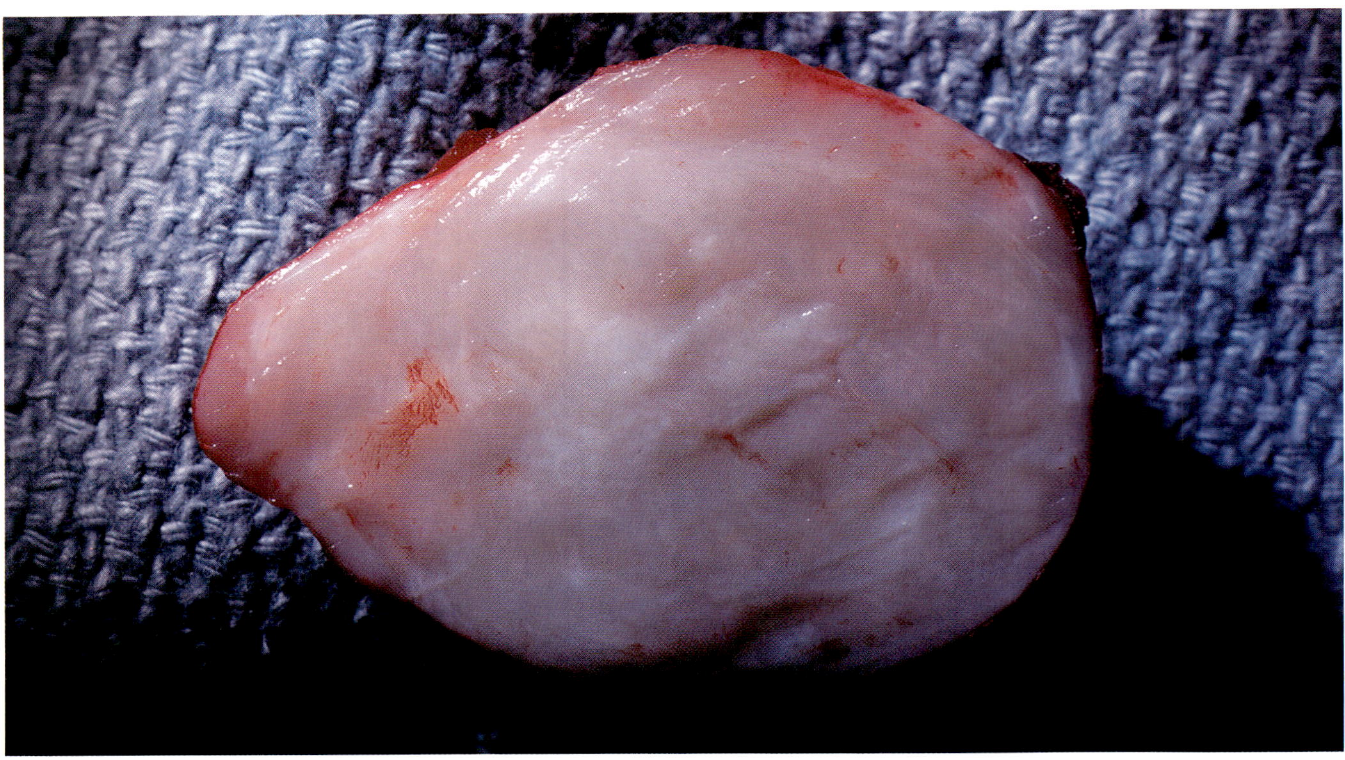

FIGURE 10–42 (ABOVE) Gross photograph from the excision specimen from the patient in Figures 10–39, 10–40, and 10–41 shows an encapsulated tumor with a white fibrous and slippery surface.

FIGURE 10–43 (BELOW) Photomicrograph from the patient in Figure 10–39 shows a tightly packed spindle-cell tumor with wavey nuclei and areas of palisading (Verocay bodies). Areas of loose fibrous tissue (Antoni A) and more cellular areas (Antoni B) provide criteria for a diagnosis of schwannoma. (hematoxylin and eosin, × 40)

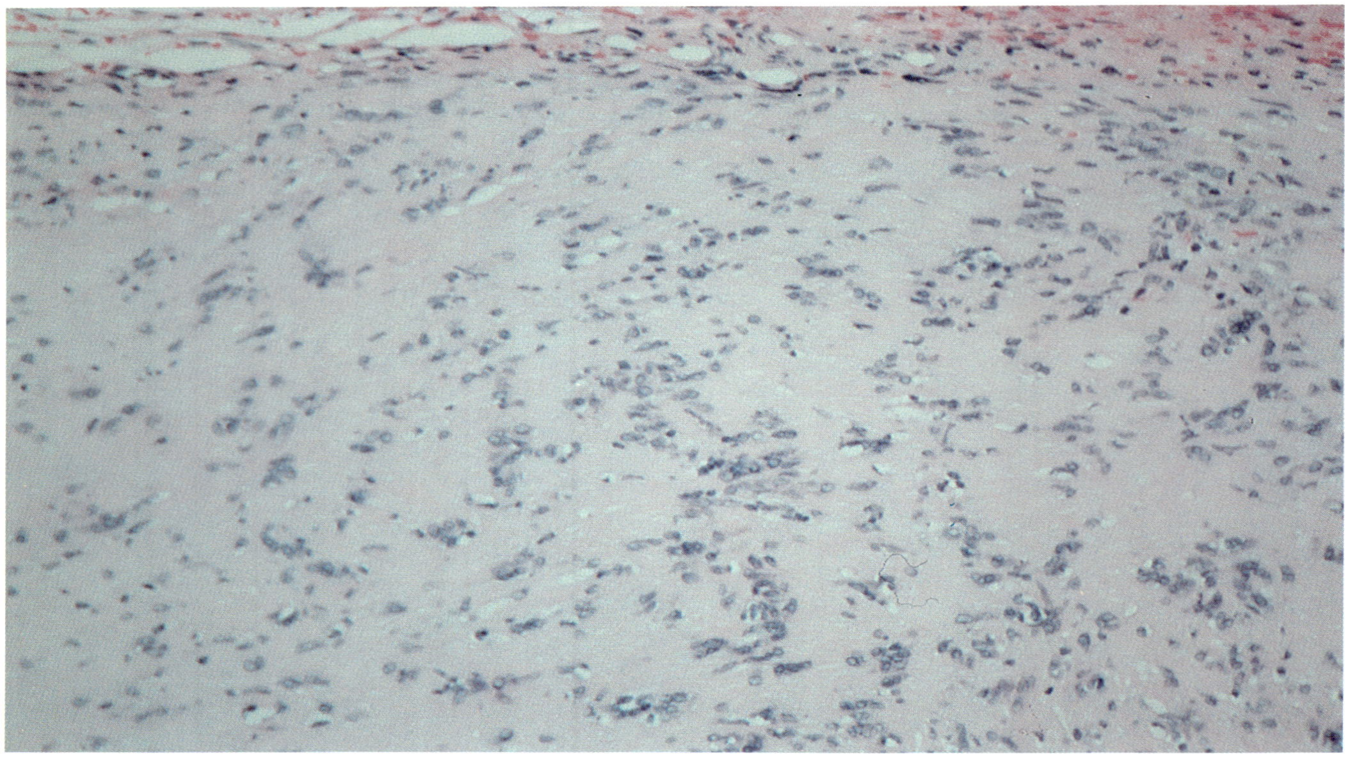

tures. Clinical presentation is determined by location. Optic canal meningiomas compress the optic nerve and lead to early and profound visual deficits.[63] Radiologic findings may show well-defined soft-tissue lesions with hyperostosis when bone is involved. The optic nerve may show fusiform swelling, diffuse thickening, or globular enlargement.[64] Fine needle aspiration has been reported to effectively diagnose orbital meningiomas in 10 cases.[3, 8, 65–68] Aspiration biopsy is appropriate for unresectable meningiomas if a tissue diagnosis is required for radiation therapy. Aspiration under CT guidance is helpful to place the needle in orbital apex and posterior meningiomas (Figures 10–44 and 10–45). Smears show oval and round cells organized in tight clusters and occasional whorls. The nuclei may have intranuclear inclusions (Figure 10–46). Rarely psamomma bodies are seen.

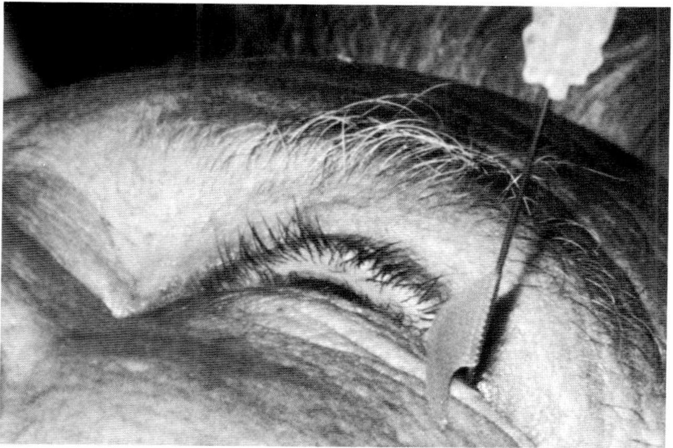

Figure 10–44 Fine needle is in place during the aspiration of the orbital mass. Sterile tape marks computed depth for entry into the tumor. Fine needle aspiration performed by Drs. Lynn Gordon and Howard Krauss, Jules Stein Eye Institute.

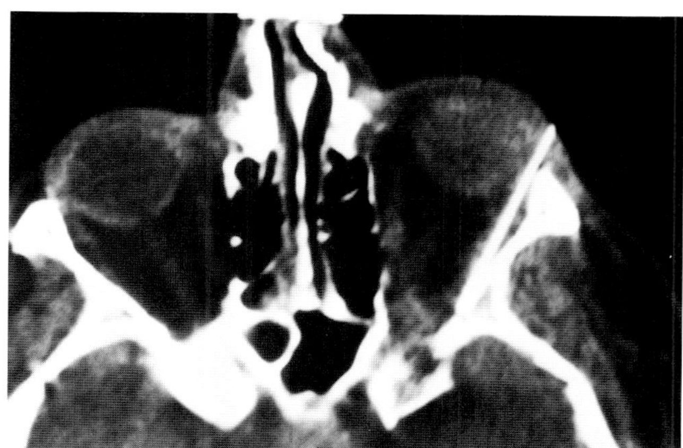

Figure 10–45 CT scan from the patient in Figure 10–44 shows a tumor at the orbital apex and confirms the placement of the needle.

CYSTIC AND VASCULAR LESIONS

Cavernous Hemangioma

Cavernous hemangiomas are well-encapsulated tumors found behind the eye within the boundaries of the rectus muscles (intraconal) and composed of very large vascular channels. The vessel walls contain smooth muscle and there is fibrous tissue in the trabeculae that separates the vessels. Cavernous hemangiomas produce slowly progressive proptosis and indentation of the posterior portion of the globe (Figure 10–47). Radiologic studies show a rounded mass with smooth contours (Figure 10–48). Echography shows a lesion with characteristic high internal reflectivity (highly echogenic). One would prefer not to aspirate this lesion because of the risk of orbital hemorrhage. However, at least four fine needle biopsies of cavernous hemangiomas have been performed without complications.[3,8,69] In every one of these cases, the fine needle aspiration revealed only the blood harbored in the large cavernous spaces (Figure 10–49).

Hemangiopericytoma

Hemangiopericytoma is a highly vascular tumor, presumably arising from pericytes, with a characteristic staghorn appearance of vessels. It occurs at a median age of 42, with twice the frequency in men as women.[70] The presenting findings of the tumor include proptosis, palpable mass, pain, and diplopia of about three years' duration. Fine needle aspiration reveals spindle and oval cells with occasional branched vessels (Figure 10–50).

Mucocele

Mucoceles are tumors composed of mucous debris and which are caused by obstruction of the ostia of the sinuses. Mucus is secreted by entrapped epithelium and enlarges the sinuses. The frontal sinus is the most common site of origin for orbital mucoceles, but they may also occur in the ethmoid, sphenoid, and maxillary sinuses.[71–75] Radiologically, the mucocele is recognizable as

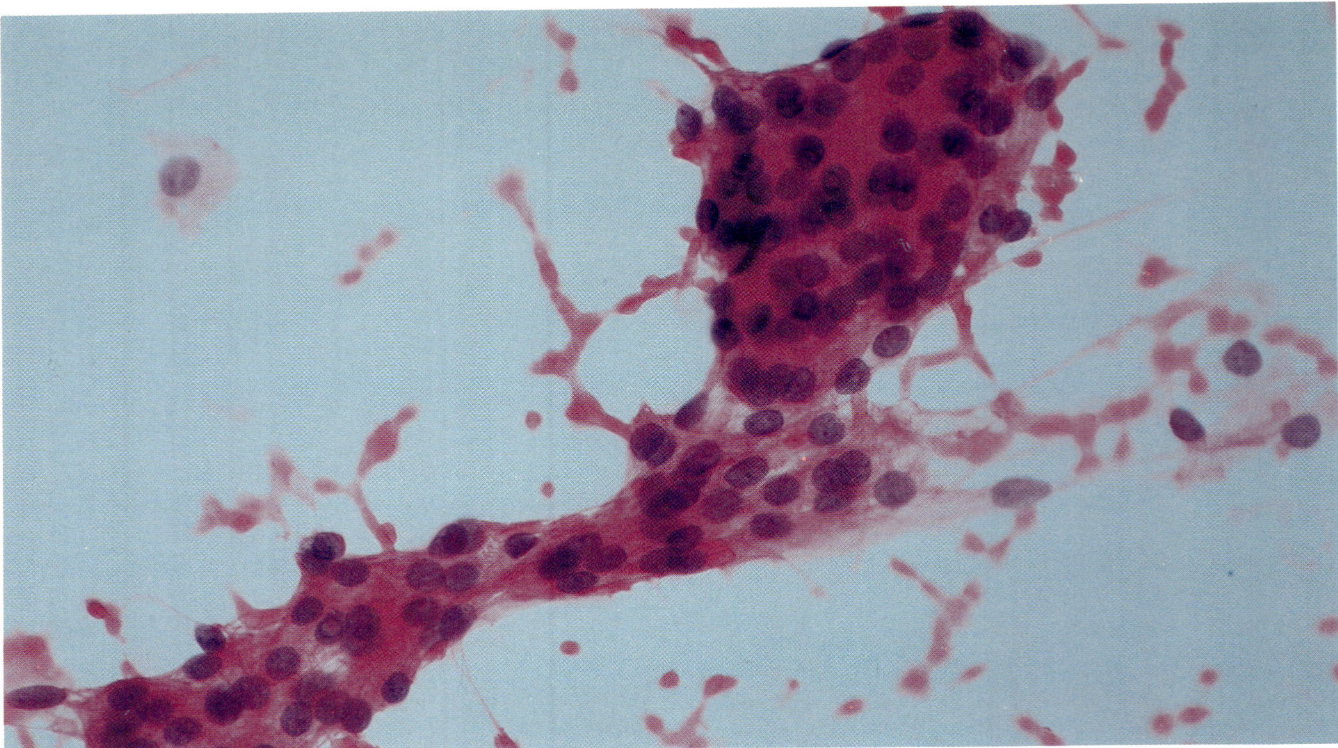

FIGURE 10–46 Photomicrograph of the aspirate for the patient in Figures 10–44 and 10–45 shows sheets of tightly clumped benign cells with regular round nuclei and small nucleoli. Occasional intranuclear inclusions are seen.

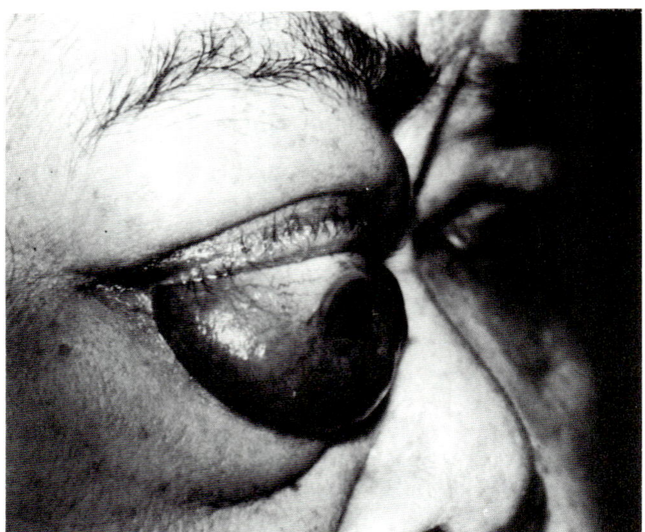

FIGURE 10–47 A 74-year-old female presented with a four-year history of progressive proptosis and swelling of the orbit.

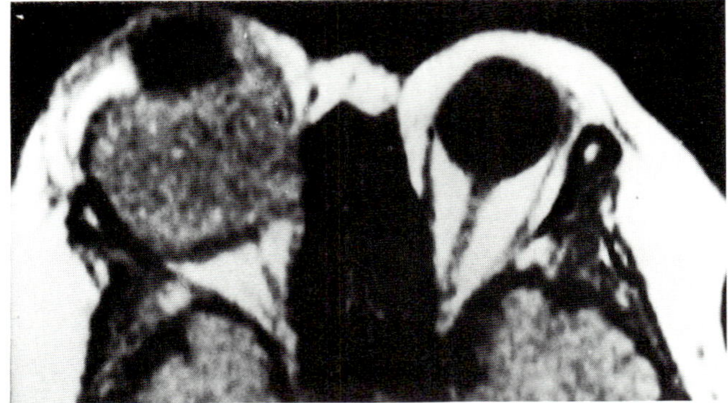

FIGURE 10–48 Magnetic resonance imaging of the patient in Figure 10–47 shows a lobulated mass filling the orbit and indenting and displacing the eye forward.

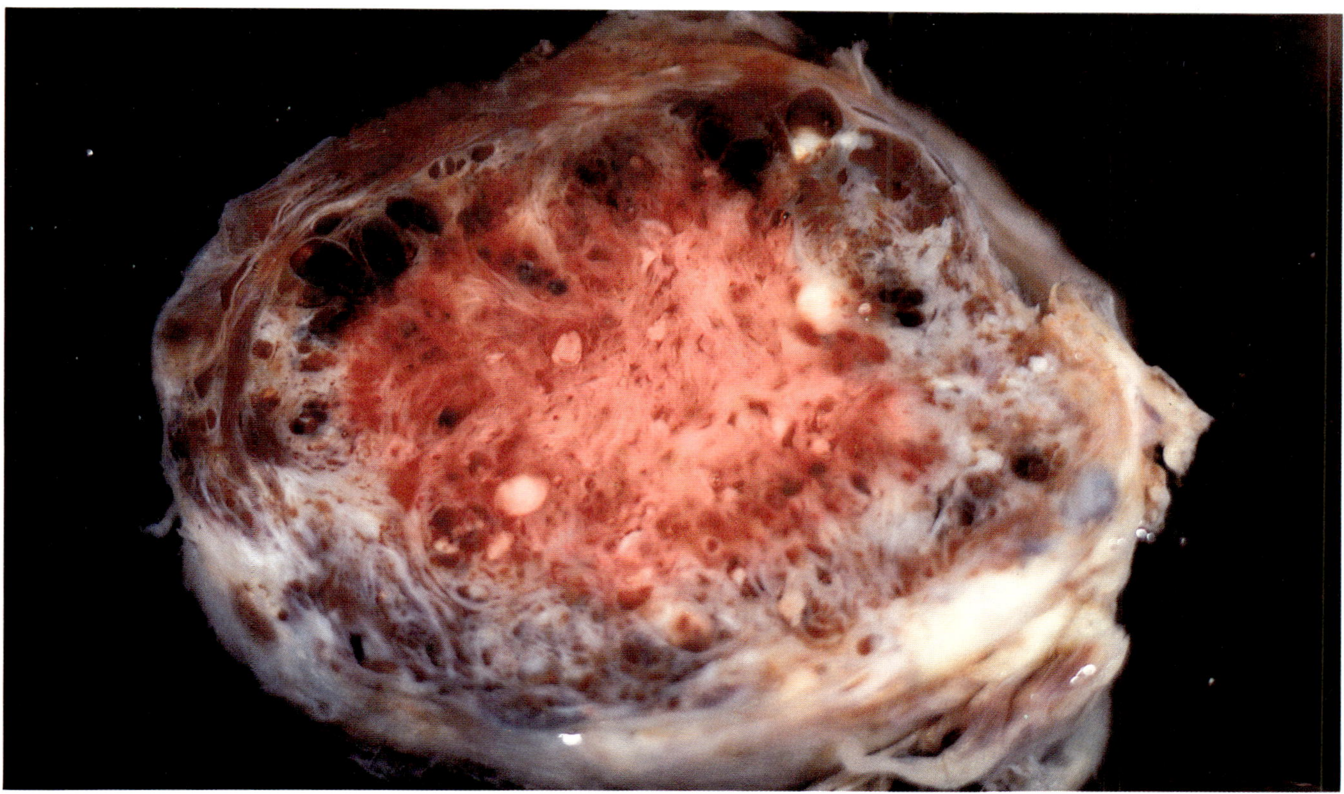

FIGURE 10–49 Gross photograph of the resected specimen from the patient in Figures 10–47 and 10–48 shows numerous dilated and enlarged channels peripherally and a central area of organized thrombosis with calcification.

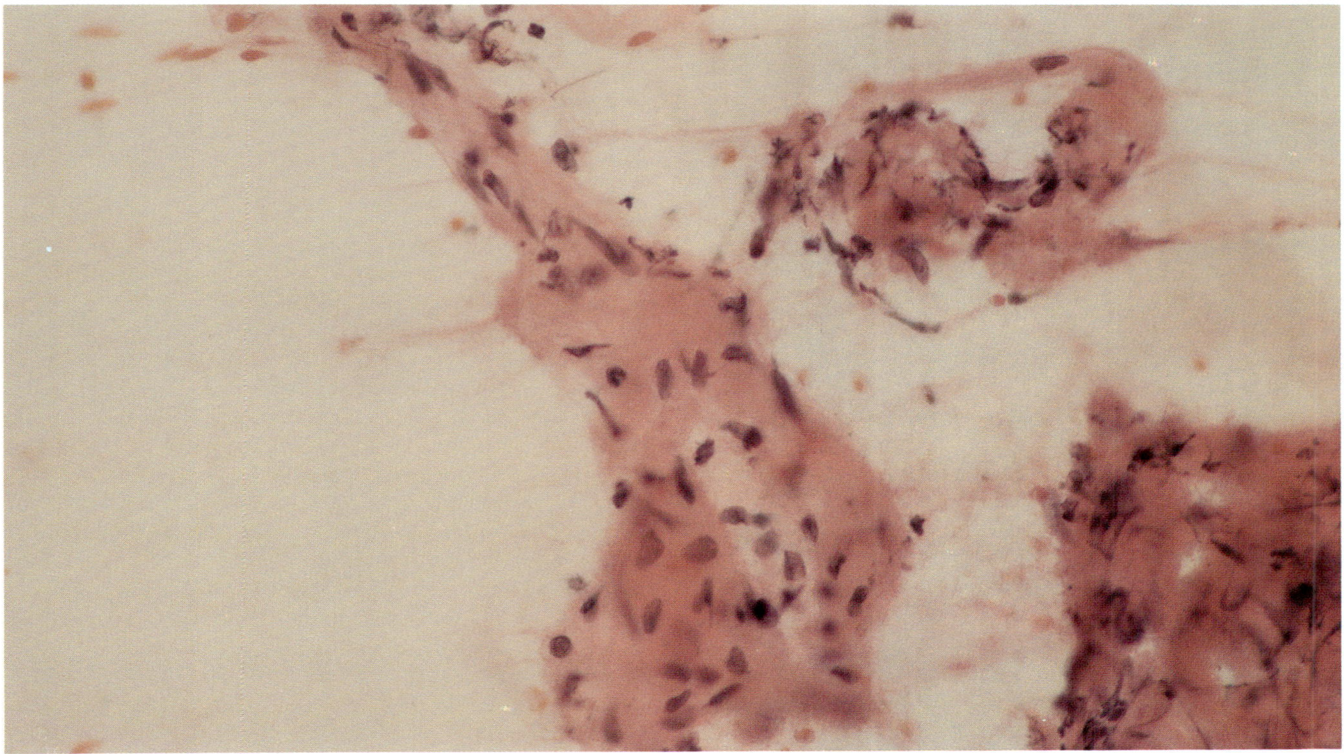

FIGURE 10–50 Smears of a fine needle aspiration biopsy show spindle and oval cells that form branched vessels (hemangiopericytoma).

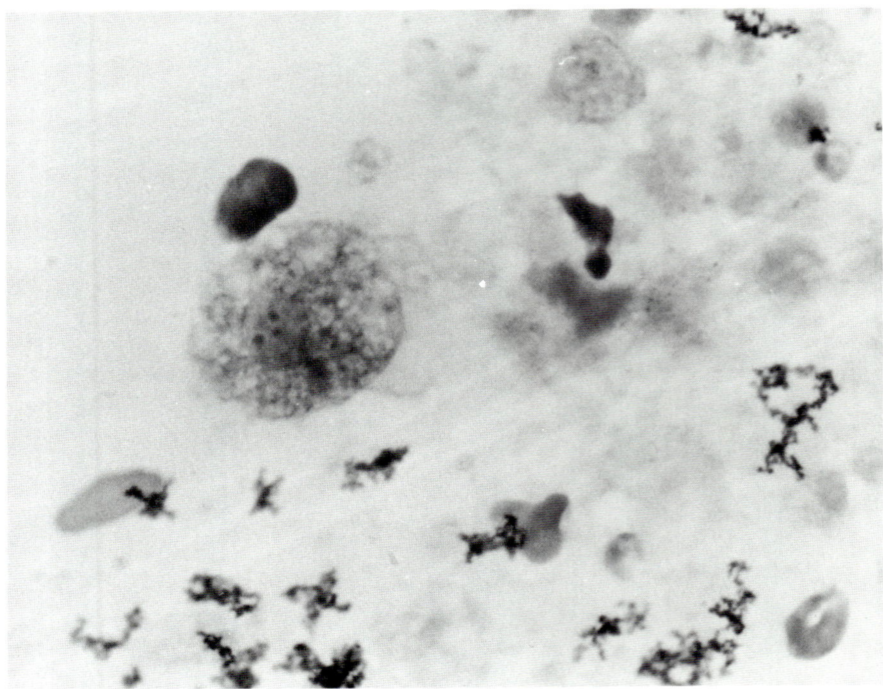

FIGURE 10–51 Smears from fine needle aspiration show large vacuolated cells with small nuclei. A background of amorphous material is present (mucocele). (hematoxylin and eosin, × 400)

a lucent mass with smooth scalloped borders expanding the sinus and destroying sinus septae around the orbit. Fine needle aspiration reveals a large amount of mucoid material with occasional vacuolated macrophages (mucophages) (Figure 10–51). In general, fine needle aspiration is done when an underlying neoplasm is suspected.

REFERENCES

1. Krohel GB, Tobin D, Chavis RM. Inaccuracy of orbital fine needle aspiration biopsy. Ophthalmology 1984[Suppl.]; 91:83.

2. Czerniak B, Woyke S, Daniel B, Krzysztolik Z, Koss LG. Diagnosis of orbital tumors by aspiration biopsy guided by computerized tomography. Cancer 1984;54:2385–2389.

3. Glasgow BJ, Layfield LJ. Fine-needle aspiration biopsy of orbital and periorbital masses. Diagn Cytopathol 1991; 7:132–141.

4. Frayer WC, Enterline HT. Embryonal rhabdomyosarcoma of the orbit in children and young adults. Arch Ophthalmol 1959;62:203–210.

5. Kirk RC, Zimmerman LE. Rhabdomyosarcoma of the orbit. Arch Ophthalmol 1969;81:559–564.

6. Yanoff M, Fine BS. Ocular pathology. A text and atlas. Philadelphia: J.B. Lippincott, 1989:532–534.

7. Knowles DM, Jakobiec FA, Potter G, Jones IS. Ophthalmic striated muscle neoplasms. A clinico-pathologic review. Surv Ophthalmol 1976;21:219–261.

8. Kennerdell JS, Slamovits TL, Dekker A, Johnson BL. Orbital fine-needle aspiration biopsy. Am J Ophthalmol 1985;99:547–551.

9. Zajdela A, Vielh P, Schlienger P, Haye C. Fine-needle aspiration cytology of 292 palpable orbital and eyelid tumors. Am J Clin Pathol 1990;93:100–104.

10. de Jong ASH, van Kessel-van Vark M, van Heerde P. Fine needle aspiration biopsy diagnosis of rhabdomyosarcoma. Acta Cytol 1987;31:573–577.

11. Wharam M, Beltangady M, Hays D, Heyn R, Ragab A, et al. Localized orbital rhabdomyosarcoma. An interim report of the intergroup rhabdomyosarcoma study committee. Ophthalmology 1987;94:251–254.

12. Jakobiec FA, McLean I, Font R. Clinicopathologic characteristics of orbital lymphoid hyperplasia. Ophthalmology 1979;86:948–966.

13. Knowles DM II, Jakobiec FA. Orbital lymphoid neoplasms, a clinicopathologic study of 60 patients. Cancer 1980;46:576–589.

14. McNally L, Jakobiec FA, Knowles DM II. Clinical, morphologic, immunophenotypic, and molecular genetic analysis of bilateral ocular adnexal lymphoid neoplasms in 17 patients. Am J Ophthalmol 1987;103:555–568.

15. Chen P, Liu JH, Lin SH, Hsu WM, Kao SC. Rearrangements of immunoglobulin gene and oncogenes in ocular adnexal pseudolymphoma. Current Eye Res 1991; 10:493–500.

16. Southern EM. Detection of specific sequences among DNA fragments separated by gel electrophoresis. J Mol Biol 1975;98:503–517.

17. Wahl GM, Stern M, Stark GR. Efficient transfer of large DNA fragments from agarose gels to diazobenzyloxmethyl-paper and rapid hybridization by using dextran sufate. Proc Natl Acad Sci USA 1979;76:3683–3687.

18. Rigby PW, Dieckman M, Rhodes C, Berg P. Labeling deoxyribonucleic acid to high specific activity in vitro by nick translation with DNA polymerase I. J Mol Biol 1977;113:237–251.

19. Yanagi Y, Yoshikai Y, Leggett K, Clark SP, Aleksander I, et al. A human T cell-specific cDNA clone encodes a protein having extensive homology to immunoglobulin chains. Nature 1984;308:145–149.

20. Flug F, Pelicci PG, Bonetti F, et al. T-cell receptor gene rearrangement as markers of lineage and clonality in T-cell neoplasia. Proc Natl Acad Sci USA, 1985;82:3460–3464.

21. Knowles DM, Athan E, Ubriaco A, McNally L, Inghirami G, et al. Extranodal noncutaneous lymphoid hyperplasias represent a continuous spectrum of B-cell neoplasia: demonstration by molecular genetic analysis. Blood 1989;73:1635–1645.

22. Levitt S, Cheng L, DuPuis MH, Layfield LJ. Fine needle aspiration diagnosis of malignant lymphoma with confirmation by immunoperoxidase staining. Acta Cytol 1985; 29:895–902.

23. Knowles DM II, Jakobiec FA. Cell marker analysis of extranodal lymphoid infiltrates: to what extent does the determination of mono- or polyclonality resolve the diagnostic dilemma of malignant lymphoma v pseudolymphoma in an extranodal site? Semin Diagn Pathol 1985;2:163–168.

24. Nikaido H, Mishima H, Kiuchi Y, Nanba K. Primary orbital malignant lymphoma: a clinicopathologic study of 17 cases. Albrecht von Graefes Arch Klin Exp Ophthalmol 1991;229:206–209.

25. Jakobiec FA, Jones IS. Introduction to ultrastructure, inflammation, and neoplasia. In: Jones IS, Jakobiec FA, eds. Diseases of the orbit. Hagerstown, MD: Harper and Row, 1979:145–179.

26. Goldberg L, Tac A, Romano P. Severe exophthalmos secondary to orbital myopathy not due to Graves' disease. Br J Ophthalmol 1982;66:392–395.

27. Kennerdell JS, Dresner SC. The nonspecific orbital inflammatory syndromes. Surv Ophthalmol 1984;29:93–103.

28. Font RL, Gamel JW. Epithelial tumors of the lacrimal gland: an analysis of 265 cases. In: Jakobiec FA, ed. Ocular and adnexal tumors. Birmingham, AL: Aesculapius Publishing, 1978:787–805.

29. Ni C, Cheng SC, Dryja TP, Cheng TY. Lacrimal gland tumors. A clinico-pathological analysis of 160 cases. Int Ophthalmol Clin 1982;22:99–120.

30. Ashton N. Epithelial tumors of the lacrimal gland. Mod Probl Ophthalmol 1975;14:306–311.

31. Zimmerman LE, Sanders TE, Ackerman LV. Epithelial tumors of the lacrimal gland: prognostic and therapeutic significance of histologic types. Int Ophthalmol Clin 1962;2:337–367.

32. Glasgow BJ, Brown HH, Zaragoza AM, Foos RY. Quantitation of tumor seeding from fine needle aspiration of ocular melanomas. Am J Ophthalmol 1988;105:538–546.

33. Layfield LJ, Tan P, Glasgow BJ. Fine needle aspiration cytology of primary salivary gland lesions. Arch Pathol Lab Med 1987;111:346–353.

34. Font RL, Gamel JW. Adenoid cystic carcinoma of the lacrimal gland. A clinicopathologic study of 79 cases. In: Nicholson DH, ed. Ocular pathology update. New York: Masson Publishing USA, 1980:277–283.

35. Perzin K, Gullane P, Clairmont A. Adenoid cystic carcinoma arising in salivary glands: a correlation of histologic features and clinical course. Cancer 1978:42:265–275.

36. Lloyd GAS. Lacrimal gland tumours: the role of CT and conventional radiology. Br J Radiol 1981;54:1034–1038.

37. Gamel JW, Font RL. Adenoid cystic carcinoma of the lacrimal gland: the clinical significance of a basaloid histologic pattern. Hum Pathol 1982;13:219–225.

38. Lee DA, Campbell RJ, Waller RR, Ilstrup DM. A clinicopathologic study of primary adenoid cystic carcinoma of the lacrimal gland. Ophthalmology 1985;92:128–134.

39. Henderson JW, Nealt RW. En bloc removal of intrinsic neoplasms of the lacrimal gland. Am J Ophthalmol 1976;82:905–909.

40. Marsh JL, Wise DM, Smith M, Schwartz H. Lacrimal gland adenoid cystic carcinoma: intracranial and extracranial en bloc resection. Plast Reconst Surg 1981;68:577–585.

41. Layfield LJ, Glasgow BJ, DuPuis MH. Fine-needle aspiration of lymphadenopathy of suspected infectious etiology. Arch Pathol Lab Med 1985;109:810–812.

42. Font RL. Eyelids and lacrimal drainage system. In: Spencer WH, ed. Ophthalmic pathology, an atlas and textbook. Philadelphia: W.B. Saunders, 1986;3:2266–2268.

43. Arora R, Rewari R, Betheria SM. Fine needle aspiration cytology of eyelid lesions. Acta Cytol 1990;34:227–232.

44. Henderson JW. Orbital tumors, 2nd ed. New York: Brian C. Decker (Thieme-Stratton), 1980.

45. Rootman J, Lapointe JS. Structural lesions. In: Rootman J, ed. Diseases of the orbit. Philadelphia: J.B. Lippincott, 1988:481–488.

46. Westman-Naeser S, Naeser P. Tumours of the orbit diagnosed by fine-needle biopsy. Acta Ophthalmol 1978;56: 969–976.

47. Kennerdell JS, Dekker A, Johnson BL, Dubois PJ. Fine needle aspiration biopsy: its use in orbital tumors. Arch Ophthalmol 1979;97:1315–1317.

48. Straatsma BR. Ocular manifestation of Wegener's granulomatosis. Am J Ophthalmol 1957;44:789–799.

49. Baghdassarian SA, Shammas HF. Eosinophilic granuloma of orbit. Ann Ophthalmol 1977;9:1247–1251.

50. Jakobiec FA, Trokel SL, Aron-Rosa D, Iwamoto T, Doyon D. Localized eosinophilic granuloma (Langerhans' cell histiocytosis) of the orbital frontal bone. Arch Ophthalmol 1980;98:1814–1820.

51. Layfield LJ, Bhuta S. Fine-needle aspiration cytology of histiocytosis X. Diagn Cytopathol 1988;4:140–143.

52. Johnson LN, Krohel GB, Yeon EB, Parnes SM. Sinus tumors invading the orbit. Ophthalmology 1984;91:209–217.

53. Shields JA, Font RL. Meibomian gland carcinoma presenting as a lacrimal gland tumor. Arch Ophthalmol 1974;92:304–306.

54. Shields JA, Font RL. Meibomian gland carcinoma presenting as a lacrimal gland tumor. Arch Ophthalmol 1974; 92:304–306.

55. Das KK, Das J, Natarajan R, Chachra KL, Chacchra KL, et al. Meibomian gland carcinoma initially identified by cytology. Diagn Cytopathol 1986;2:154–156.

56. Boniuk M, Zimmerman LE. Sebaceous gland carcinoma of the eyelid, eyebrow, caruncle and orbit. Trans Am Acad Ophthalmol Otolaryngol 1968;72:619–642.

57. Rao NA, Hidayat AA, McLean IW, Zimmerman LE. Sebaceous gland carcinoma of the ocular adnexa: a clinicopathologic study of 104 cases with five year follow-up data. Hum Pathol 1982;13:113–122.

58. Aurora AL, Blodi FC. Lesions of the eyelids. A clinicopathologic study. Surv Ophthalmol 1970;15:94–104.

59. Goldberg RA, Rootman J. Clinical characteristics of metastatic orbital tumors. Ophthalmology 1990;97:620–624.

60. Kopelman JE. Shorr N. A case of prostatic carcinoma

metastatic to the orbit diagnosed by fine-needle aspiration biopsy and immunoperoxidase staining for prostatic specific antigen. Ophthalmic Surg 1987;18:599–603.

61. Izumi AK, Rosato RE, Wood MG. Von Recklinghausen's disease associated with multiple neurilemmomas. Arch Dermatol 1971;104:172–176.

62. Rootman J, Robertson WD. Tumors. In: Rootman J, ed. Diseases of the orbit. Philadelphia: J.B Lippincott, 1988: 319–321.

63. Wilson WB. Meningiomas of the anterior visual system. Surv Ophthalmol 1981;26:109–127.

64. Rootman J, Robertson WD. Tumors. In: Rootman J, ed. Diseases of the orbit. Philadelphia: J.B. Lippincott, 1988: 293–305.

65. Meyer E, Malberger E, Gdal-on M, Zonis S. Fineneedle aspiration of orbital lesions. Ann Ophthalmol 1983;15: 635–638.

66. Czerniak B, Woyke S, Daniel B Krzysztolik Z, Koss LG. Diagnosis of orbital tumors by aspiration biopsy guided by computerized tomography. Cancer 1984;54:2385–2389.

67. Zajdela A, de Maublanc MA, Schlienger P, Haye C. Cytologic diagnosis of orbital and periorbital palpable tumors using fine-needle sampling without aspiration. Diagn Cytopathol 1986;2:17–20.

68. Cristallini EG, Bolis GB, Ottaviano P. Fine-needle aspiration biopsy of orbital meningioma. Acta Cytol 1990;34: 236–238.

69. Meyer E, Malberger E, Gdal-on M, Zonis S. Fineneedle aspiration of orbital lesions. Ann Ophthalmol 1983; 15:635–638.

70. Croxatto JO, Font RL. Hemangiopericytoma of the orbit: a clinicopathologic study of 30 cases. Hum Pathol 1982;13:210–218.

71. Montogomery WW. Mucocele of the maxillary sinus causing enophthalmos. Eye Ear Nose Throat J 1964;43:41.

72. Alberti PW, Marshall HF, Black JI. Fronto-ethmoidal mucocele as a cause of unilateral proptosis. Br J Ophthalmol 1968;52:833.

73. Guerry R, Smith J. Paranasal sinus carcinoma causing orbital mucocele. Am J Ophthalmol 1975;80:943–946.

74. Johnson LN, Hepler RS, Yee RD, Batzdorf U. Sphenoid sinus mucocele (anterior clinoid variant) mimicking diabetic ophthalmoplegia and retrobulbar neuritis. Am J Ophthalmol 1986;102:111–115.

75. Stanton MB. Sphenoid sinus mucocele. Am J Ophthalmol 1970;70:991–994.

Index

References to figures are in bold italic.